LAND SURVEYING MATHEMATICS SIMPLIFIED

PAUL L. GAY

Land Surveying Mathematics Simplified

© Copyright 2017 Paul L. Gay

ISBN #: 978-1-387-41776-6

Rev. 5

To my wife Viola, a tireless researcher and a surveyor in her own right, and to my mother Elinor who encouraged me to become a surveyor.

A Note on My Book Practical Boundary Surveying.

This book was written to make it readable by the widest possible audience. Should the reader be interested in a more detailed treatment of the subject my book *Practical Boundary Surveying, Legal and Technical Principles*, Paul Gay, Springer Science Publishing 2015 goes into considerably more detail. It also has many examples. The book contains an appendix with the most commonly used equations and sample calculations used by surveyors making it useful to those requiring a more in-depth treatment of the subject, including students of land surveying at the college level.

About the Author

Paul L. Gay is a professional land surveyor and attorney. He has been in private practice for more than 35 years. His primary interest in surveying has been boundary surveying and boundary dispute resolution. Mr. Gay has served as an expert witness in trials concerning boundaries from Massachusetts and Rhode Island to Alaska.

Mr. Gay's first book on land surveying was entitled *Fundamentals of Boundary Surveying, How Boundaries are Established*, published by Professional Surveyor Publishing Co., Inc., in 2002. His most recent hard cover book is entitled *Practical Boundary Surveying, Legal and Technical Principles*, published by Springer Science Publishing in 2015. Mr. Gay also wrote *Survey*, a computer program which performs surveying coordinate geometry calculations. He also wrote *Tide*, a computer program which calculates tidal heights for most coastal locations in the U.S. Other publications include *A history of Gray's Mill, Basic Principles of 19th Century Water Power* and *Sediment Transport around Gooseberry Neck*. Mr. Gay holds a B.S. degree in Sociology from the University of Massachusetts, an ScM degree in Geological Sciences from Brown University and a J.D. degree from New England Law.

Other Books by the Author

Practical Boundary Surveying, Legal and Technical Principles, Paul L. Gay, Springer Science Publishing in 2015.

Land Surveying Simplified, Paul L. Gay, 2016. Available on Lulu as an eBook or Paperback. *Land Surveying Simplified* is designed to be read along with this book.

A History of Gray's Mill, Paul L. Gay 2016. A short history of a grist mill which has been in operation for over 360 years, located in Westport, MA. Available on Lulu in Paperback.

Sidney, A Short Story, Paul L. Gay, 2016. A short humorous fiction about a wind turbine developer during the energy crisis of the late 1970s. Available as an eBook free of charge on Lulu.

Introduction

This book is designed to provide the reader with an understanding of basic mathematics commonly used by land surveyors. Because the focus of this book is to provide readers with a "basic" introduction to the subject in roughly 100 pages, it was necessary to limit the scope of the coverage somewhat. This book is not intended to provide a comprehensive treatment of the subject such as might be found in a college course or in a textbook intended to prepare applicants for taking land surveyor licensing examinations.

This book is intended to be a companion book to *Land Surveying Simplified*, which is available on Lulu as and eBook or soft cover. Land Surveying Simplified provides a largely non-mathematical overview of how boundary surveys are performed. Following in the footsteps of Land Surveying Simplified, this book is primarily concerned with the mathematics which boundary surveyors use. It is not intended to be a guide to calculating topographic surveying, control surveying or construction surveying, although all of the principles discussed here can be applied to these disciplines. The presentation in this book is intended to be as simplified as possible and the author has attempted to provide clearly worded explanations of each subject. Although a background in algebra, geometry, trigonometry and statistics would be helpful, none of these is strictly necessary in order for the reader to work through these materials.

Many examples are given to help the reader understand the concepts. Computers and software have made surveying calculations very quick and easy. This does not relieve a land surveyor from understanding the principles underlying a specific task or calculation. The material presented in this book is intended to provide the reader with a basic understanding of the mathematics used by boundary surveyors. We will not spend much time on abstractions and things that are interesting and nice to know, but really not that essential to a working knowledge of the subject.

So, let's get started…

Table of Contents

Angles and Trigonometry

What is an angle? **Angles** describe the relationship between two intersecting lines. In the United States land surveyors customarily use the **sexigesimal system** for describing angles.   In the sexigesimal system angles are measured using **degrees**, **minutes** and **seconds**. The standard notation for degrees is a small circle in superscript after the number. Ten degrees would be written as 10°. The standard notation for minutes is a short tick mark in superscript after the number (an apostrophe). 10 minutes would be written as 10'.   Seconds are noted with two tick marks after the number (a quote). 10 seconds would be written as 10".

The sexigesimal system is as follows:

1. There are 360 degrees (360°) in a full circle.
2. Each degree contains 60 minutes (60').
3. Each minute contains 60 seconds (60").

As you can see in **Figure 1**, angles are always measured from a reference line. You can also see from the figure that angles can be measured to the right (clockwise) or to the left (counterclockwise).

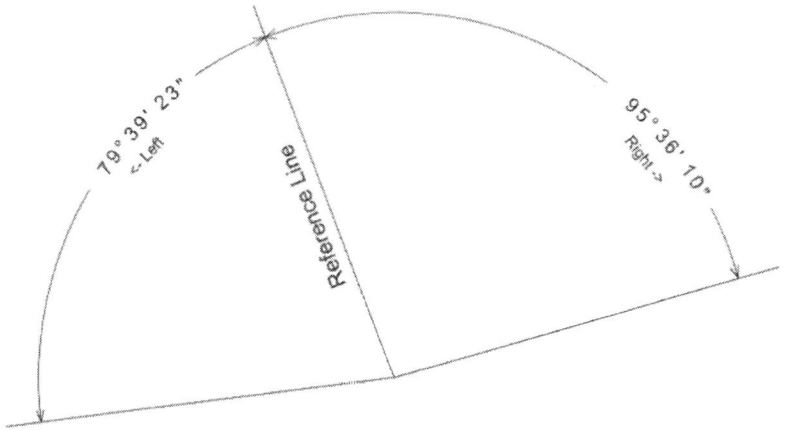

79° 39' 23"
←Left

Reference Line

95° 36' 10"
Right →

Figure 1 – Measurement of angles.

The reader is encouraged to do their own calculation to prove that there are 21,600 minutes in a full circle (360°), and that there are 1,296,000 seconds in a full circle. You may also want to do the calculation showing that there are 3,600 seconds in a degree.

When writing angles, it is customary to limit the maximum number of minutes and seconds to 59. For example, assume we were to write the following angle:

23° 60'

We have learned that there are 60 minutes in a degree, so we would subtract 60 from the minutes column and add one degree to the degrees column. The angle would be written as:

24° 00'

The same procedure would be used if the seconds were 60. Recall that one minute contains 60 seconds so 60 would be subtracted from the seconds column and one minute added to the minutes column. The following angle:

123° 44' 60"

Would be written as:

123° 45' 00"

Even though a circle cannot have more than 360 degrees, we will see that when adding a large number of angles together the sum may well exceed 360 so there is actually no limit to the number of degrees that can be written.

Adding Angles

Adding angles is straightforward. The rule is to start at the rightmost column and work left. For example, assume that you want to add the following two angles:

120° 11'
12° 37'

You would start by adding 37 minutes to 11 minutes, then move on to adding the degrees. Notice that the two angles in our example have no seconds. If you wanted to add the following two angles you would start at the right by adding seconds first (02" + 55"), then the minutes (36' + 21") and last the degrees (45° + 22°).

45° 21' 55"
22° 36' 02"

Let us take a look at adding some angles.

Example 1

Add 4° 23' 12" and 14° 05' 11"

First we add the seconds: 12" + 11" = 23".

Nest we add the minutes: 23' + 5' = 28'.

Last, we add the degrees: 4° + 14° = 18°

Our answer is: 18° 28' 23"

Example 2

Let's try a more difficult example. In this example, notice that the sum of the seconds will exceed 59 so we will need to subtract 60 from the seconds column and add one minute to the minutes column.

Add the following two angles:

44° 11' 56" and 10° 14' 46"

Add the seconds: 56" + 46" = 102"

The angle is greater than 59 so subtract 60" and add 1' to the minutes.

102" – 60" = 42"

Next, add the minutes:

11' + 14' + 1' = 26'

Last, add the degrees:

44° + 10° = 54°

Our answer is 54° 26' 42"

Example 3

One more example of adding two angles.

Add 90° 49' 39" and 56° 54' 35"

Add the seconds: 35" + 39" = 74"

Subtract 60": 74" – 60" = 14"

Add the minutes: 49' + 54' + 1' = 104'

Subtract 60': 104' = 60' = 44'

Add the degrees: 90° + 56° +1° = 147°

Our answer is: 147° 44' 14"

In the above examples, the value of seconds and minutes was less than 120. It is possible that the value of minutes or seconds is substantially greater.   This often happens when adding many angles together as in the following example:

Example 4

$$59° \ 23' \ 54"$$
$$22° \ 41' \ 46"$$
$$10° \ 35' \ 44"$$
$$\underline{55° \ 37' \ 59"}$$
$$146° \ 136' \ 203'$$

The sums shown in Example 4 are simply the totals for each column. You can see that both the seconds column and minutes column the sums are greater than 59. However, subtracting 60 would still leave a number greater than 59.

In such cases we just need to recognize that the value is some multiple of 60. In the present example, 203 seconds is greater than 180 seconds (3 * 60 = 180). 203 seconds is also less than 240 seconds (4 * 60 = 240). Our determination is made simply by dividing the number by 60 and looking at the integer portion of the number. For example, 203 / 60 = 3.38. The result tells us that the seconds are evenly divisible by 3 with some remainder of seconds. So, 3 minutes = 180 seconds (3 * 60 = 180). The result is:

203" – 180" = 23"

We would then add 3 minutes to the minutes column giving a result of 139 minutes. Notice that 139 minutes is divisible by 2 so we would subtract 120 minutes and add 2 to the degrees column. The result is:

$$59° \ 23' \ 54"$$
$$22° \ 41' \ 46"$$
$$10° \ 35' \ 44"$$
$$\underline{55° \ 37' \ 59"}$$
$$148° \ 19' \ 23"$$

Adding a large number of angles is very common when it is necessary to calculate the closure of a traverse or boundary which has many angles. We will learn in a later chapter that for any closed traverse or boundary, the sum of the interior angles has to be equal to a certain specific number, which is a function of the number of angles in the traverse or boundary. So, if you want to know how precise your measurement of the traverse angles was you can easily compare the sum of the measured angles with this number.

Subtracting Angles

Subtracting angles uses an approach similar to adding angles. Again, we start from the right. Let's perform the following subtraction:

Example 5

110° 24' 37" – 22° 12' 04" = 88° 12' 33"

Or

$$\begin{array}{r} 110° \ 24' \ 37" \\ -22° \ 12' \ 04" \\ \hline 88° \ 12' \ 33" \end{array}$$

The process is straightforward because the values in each of the columns of the numbers being subtracted are less than those in the numbers being subtracted from. Let's look at a slightly more difficult example.

Example 6

146° 32' 12" - 33° 56' 47"

Again, we start at the right-hand column. Because 47 is greater than 12 we need to subtract one minute from the minutes column and add 60 seconds to the seconds column. Doing so, we now have:

146° 31' 72"

Considering the minutes column, we notice that 56 seconds is greater than 31 seconds so we must subtract 1 degree from the degrees column and add 60 minutes to the minutes column. We now have:

145° 91' 72"

So, we can now do the subtraction:

$$145° \; 91' \; 72"$$
$$- \; 33° \; 56' \; 47"$$
$$112° \; 35' \; 25"$$

Converting DMS to Decimal Degrees

So far, we have assumed that angles are always written in the sexigesimal system. However, it is sometimes convenient or necessary to express angles in decimal degrees. When using decimal degrees, there are no minutes or seconds. The value of minutes and seconds are expressed as a fraction of a degree. For example, the following angle:

57° 30' 00"

Would be written in decimal degrees as:

57.5°

We know that because there are 60 minutes in a degree, 30 minutes is exactly a half of a degree. (For simplicity we are ignoring the importance of carrying forward the precision of the original angle in our example.)

Many scientific calculators have trigonometric functions built in. For example, it may be possible to produce the values of sines, cosines and tangents of an angle. We will learn about these functions in a later chapter but for now, if you are not familiar with trigonometric functions, just note that these functions are very commonly used by surveyors.

Some scientific calculators will not accept the input of degrees-minutes-seconds (DMS) when calculating trigonometric functions. In such cases it is necessary to input the angle in decimal degrees (DEG). Fortunately, many of these calculators are able to convert degrees-minutes-seconds (DMS) to decimal degrees and to reverse the procedure. Th conversion function key(s) may be labeled as converting hours-minutes-seconds to decimal hours.

If your calculator does not do the DMS to DEG conversion, you need to know how to do it yourself. So, let us take a look at how to do the conversion.

First, a simple example. Assume we have the angle: 145° 15'. Notice that there are no seconds that we must deal with, only minutes. All we need to do is to convert the 15 minutes into decimal degrees. We know that there are 60 minutes in one degree. We simply need to know what fraction 15 is of 60 so we proceed as follows:

$$15' = \frac{15}{60} = 0.25°$$

We add our decimal fraction to the 145 degrees and our answer is 145.25°

In another example we have the following angle to convert to decimal degrees: 22° 39' 30". Starting with the seconds, we will convert the seconds into decimal minutes.

$$30" = \frac{30}{60} = 0.500 \ minutes$$

We now add the decimal minutes to the minutes and divide the result by 60 to get decimal degrees:

$$Decimal\ degrees = \frac{39' + 0.500'}{60} = 0.6583°$$

Finally, we add the decimal portion of the degrees to the degrees to get decimal degrees.

$$Decimal\ degrees = 22.6583°$$

Converting Decimal Degrees to DMS

Let's reverse the above procedure and convert decimal degrees to degrees - minutes – seconds. Let us convert 45.3767° into DMS. Notice that the decimal portion of the angle is some fraction of a degree. We know that there are 60 minutes in a degree so, in order to calculate the number of minutes we can simply multiply the decimal portion of the degrees times 60.

$$Minutes = 0.3767 * 60 = 22.6000'$$

We now have 22 minutes and a remainder of 0.6 minutes. To get seconds we must multiply the decimal portion of the minutes by 60.

$$Seconds = 0.6000 * 60 = 36''$$

Our result is 45° 22' 36".

Trigonometric Functions

Trigonometric functions enable a surveyor to calculate both angles and the lengths of lines. A commonly taught method to understand trigonometric functions makes use of a "**unit circle**" such as the one shown in **Figure 2**. A unit circle has a radius of 1 unit (the length of the radius line). It doesn't matter what the units are. They could be inches or meters or yards.

In our unit circle, we will measure angles from the horizontal axis in a counterclockwise direction. For example, in Figure 2 the angle is 40°. Notice in this figure, that we have drawn a vertical line from the end of our radius line down to the horizontal X axis. The length of this is 0.64. We call this value a **sine**. Also notice that we have a distance of 0.77 from the bottom end of the vertical line to the center of the circle. We call this value a **cosine**. On many scientific calculators Sine is abbreviated sin and cosine is abbreviated cos.

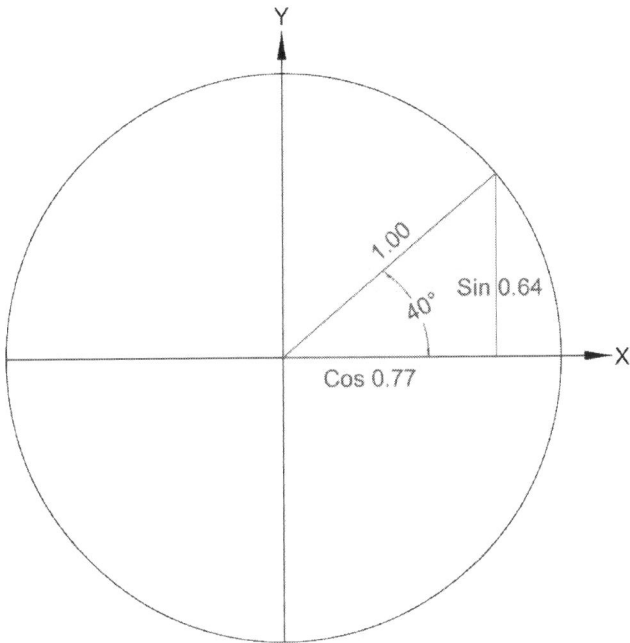

Figure 2 – Sin and Cosine.

If we rotate our unit line, the sines and cosines will change. You can see from Figure 2 that if we were to rotate our radius line counterclockwise to the left so that the angle was 90° the line would be vertical and coincident with the Y axis. In this case the sine would be 1 and the cosine would be 0. If we were to rotate the line clockwise to the right so that the line was horizontal and coincident with the X axis, the sine would be 0 and the cosine would be 1.

The unit circle tells us that both the sine and cosine will depend on the angle and will vary between 0 and 1. If the angle is 45° the sine and cosine will be the same (0.7071).

In the real world, the length of the line we are working with will usually not be exactly 1. However, we can calculate the length of any line simply by multiplying the number by the sin and cos values derived from the unit circle. For example, using the unit circle in Figure 2, if our line were actually 100.00 feet long, the X value would be:

$$100 * \cos 40° = 77 \, feet$$

And our Y value would be:

$$100 * \sin 40° = 64 \, feet$$

Because every angle has a unique sine and cosine, then every sine and cosine must have a unique angle. If the sine of 30° is 0.50, then the angle associated with the sine value of 0.50 must be 30°. When we have a sin and need to know the angle we call that value the arcsine. For example, the arcsine of 0.5 is 30°. The arccosine of 0.87 is 30°. Arcsine and arccosine are sometimes abbreviated asin and acos. Another way that arcsines and arccosines are written is using exponential notation. Arcsine is written $\sin^{-1}$ and arcos is written $\cos^{-1}$.

To summarize, sine and cosine refer to a decimal number such as 0.567. Arcsine ($\sin^{-1}$) and arccosine ($\cos^{-1}$) refer to angles.

There is one other important trigonometric function commonly used by surveyors. It is called the tangent, sometimes abbreviated tan. The tangent can be seen in Figure 3. In the unit circle diagram the tangent is always 90° to the radius line and it intersects the x axis.

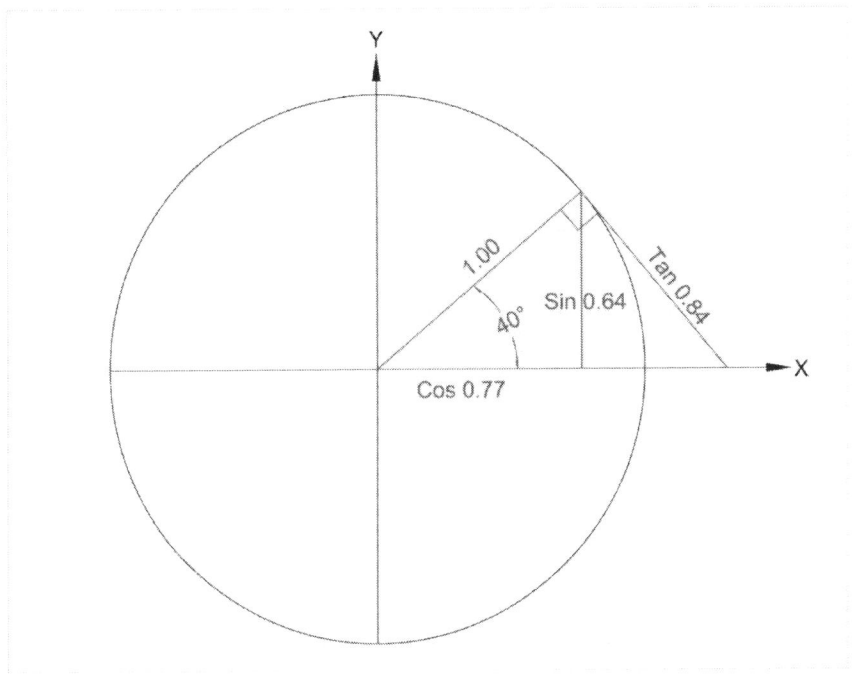

Figure 3 - Tangent

If you imagine rotating the radius line in our image clockwise so that the angle was very small, and nearly parallel to the x axis, the values of the tangent and sin would become very close to each other. If the angle was zero both the tan and sin would be zero. You can also see that if we were to rotate the radius line counterclockwise so that it approached the vertical, the tangent line

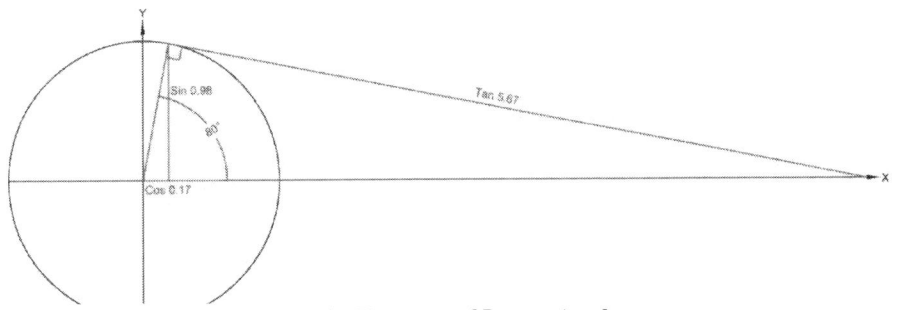

Figure 4 - Tangent of Large Angle

would become almost parallel to the x axis and its length would approach infinity. The tangent of an 80° angle (5.67) is shown in Figure 4. If the angle were actually 90°, the tangent line and the x axis would be parallel and not intersect, so the tangent of 90° is meaningless.

Mathematically, the tangent is equal to the sin divided by the cosine:

$$tan = \frac{sin}{cos}$$

As with the arcsine and arccosine there is an arctangent also noted as atan^{-1}. If you know the arctangent value, you can calculate the corresponding angle.

Right Triangles

A triangle is a geometric figure having three sides and three interior angles. In all triangles, the sum of the three interior angles must equal 180°. A special case of triangles is called a **right triangle.** In a right triangle one of the interior angles is always 90°. So, in a right triangle, the sum of the other two angles must equal 90°.

Right triangles are very commonly used in surveying. For example, right triangles form the basis of coordinate geometry. As a surveyor, you know that total stations measure distances to targets and those distances are (almost) always slope distances (in the rare instance that the target is exactly the same elevation as the total station the slope distance would be equal to the horizontal distance). Boundary measurements are always horizontal distances so surveyors need to be able to convert slope distances to horizontal distances and we will see that the right triangle makes this possible.

A typical right triangle is shown in **Figure 5**. The convention is that angles are noted with upper case letters and the length of the sides with lower case letters. In our figure, angle C is 90°. The sides of a triangle have names. The **hypotenuse** is opposite the 90° angle. It is always the longest side. The side next to angle θ is called **adjacent**. The side opposite angle θ is called **opposite**.

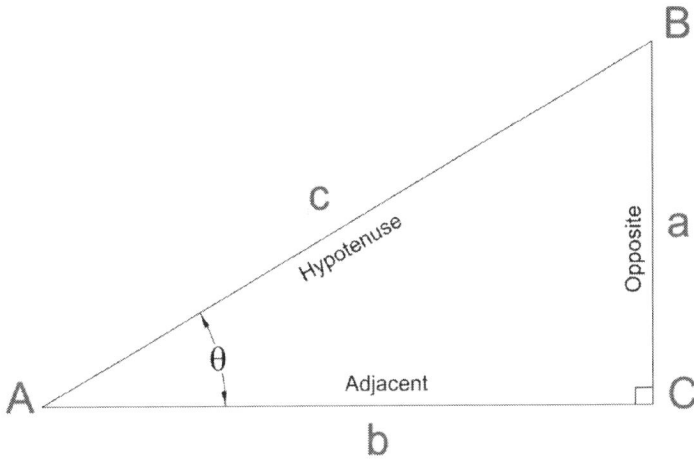

Figure 5 - Right Triangle.

The following three equations will allow you to solve any right triangle:

$$sin\ A = \frac{a}{c}$$
Equation 1

$$cos\ A = \frac{b}{c}$$
Equation 2

$$tan\ A = \frac{a}{b}$$
Equation 3

Because there are only three variables, if any two of the variables are given, a right triangle can be solved. The equations can be rearranged as necessary in order to provide the needed solution. For example, consider Equation 1. The three possible solutions are:

$$sin\ A = \frac{a}{c}$$

$$a = c * sin\ A$$

$$c = \frac{a}{sin\ A}$$

We have mentioned the common situation illustrated in **Figure 6.** When a slope distance is measured with a total station and we want to know the horizontal distance. We only need two pieces of information in order to calculate the horizontal distance: 1) the vertical angle and 2) the slope distance. Fortunately, all total stations will provide us with this information. Notice from Figure 6 and Figure 7 that the slope distance forms the hypotenuse (side c) of a right triangle. Also notice that horizontal distances are, by definition, 90° from the vertical so if the horizontal distance were the base of a right tringle, angle C would have to be 90°.

It is worth noting that total stations usually measure vertical angles from the zenith (zero is directly overhead), so we will need to subtract the vertical angle from 90° to get the angle from the horizontal plane (angle A).

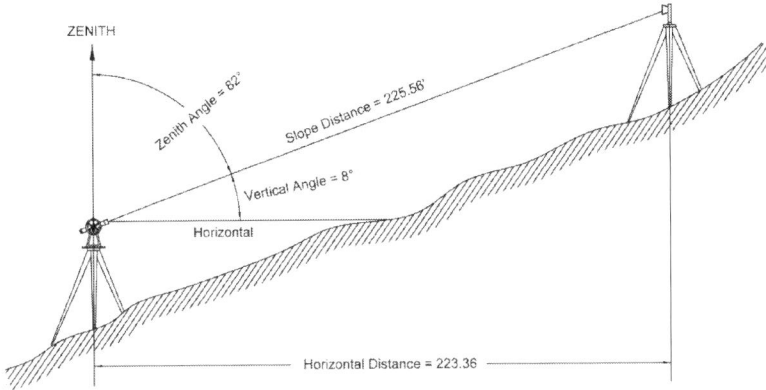

Figure 6 – Slope Distance measured with an EDM.

Example 7

Slope Distance = 225.56 and Zenith Angle = 82°

The vertical angle measured from horizontal is:

90° - 82° = 8°.

Looking at **Figure 7,** the slope distance is side c and the vertical angle, as measured from the horizontal, is angle A. So, we have side c and angle A, and we need distance b, the horizontal distance.

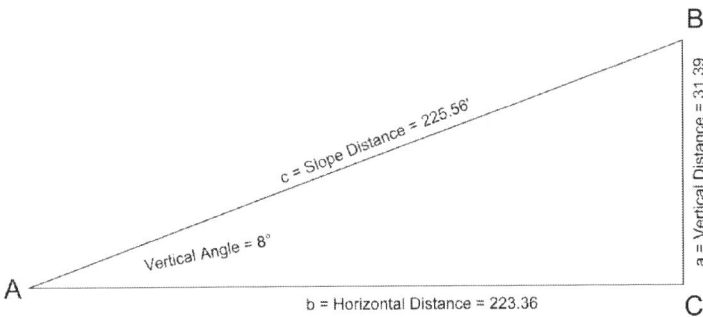

Figure 7 – Calculate Horizontal Distance from Slope Distance.

27

From the right triangle equations we select the one containing our three variables: A, b, c:

$$cos\ A = \frac{b}{c}$$

Equation 4

Because we need to solve for b we rearrange the equation as follows:

$$b = c * cos\ A$$

Substituting our values in the equation:

$$b = 225.56 * cos\ 8° \ (0.9903) = 223.36\ Feet$$

It is worth pointing out again that, in surveying, the horizontal distance is always shorter than the slope distance so if your calculation gives a horizontal distance greater than the slope distance there has to be an error in the calculation.

Example 8

Here is another example shown in Figure 8. Suppose that you need to measure the height of a shed. You measure the angle to the top of the shed (20°) and measure the distance to the shed (15.00 feet). Both the shed and total station are on level ground and the total station is 5.00' above the ground.

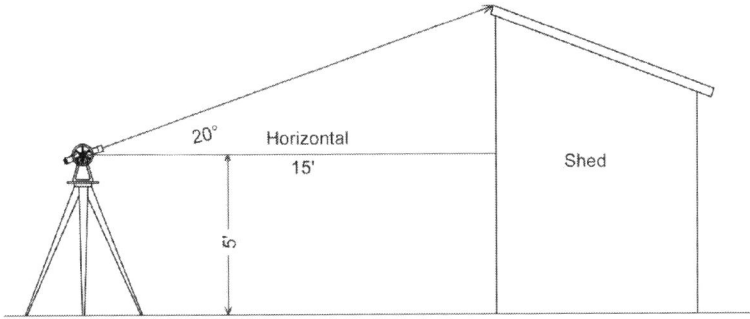

Figure 8 – Shed Example

Looking at Figure 8 the angle to the peak of the shed is angle A and the distance to the shed is side b. The height of the shed above the total station is side a. So, we have A and b and want to solve for a. We can use Equation 3.

$$tan\ A = \frac{a}{b}$$

Rearrange to solve for a.

$$a = b * \tan A$$

$$a = 15.00' * \tan(20°) = 5.46'$$

Of course, we would need to add the instrument height of 5' so the height of the shed would be 10.46'

Our last subject on right triangles is sometimes referred to as the Pythagorean theorem. It states that the square of the hypotenuse is equal to the sum of the squares of the other two sides. In mathematical notation it looks like this:

$$c^2 = a^2 + b^2 \qquad\qquad \text{Equation 5}$$

If you know the length of any two sides you can use this equation to calculate the length of the third side. Notice that this equation does not use angles as did our previous equations. Using Figure 9 as an example, assume that we want to install solar panels on the roof of the house. In order to calculate how many panels we can fit on the roof we need to know the length of the roof measured along the slope. We don't have a ladder, so we have no way to get up there to actually measure the length of the roof along the slope. However, we know that the peak of the roof is 26' high and the side wall is 10' high. We also know that the house is 40' wide and the peak is in the middle, so it is 20' from the side to the peak.

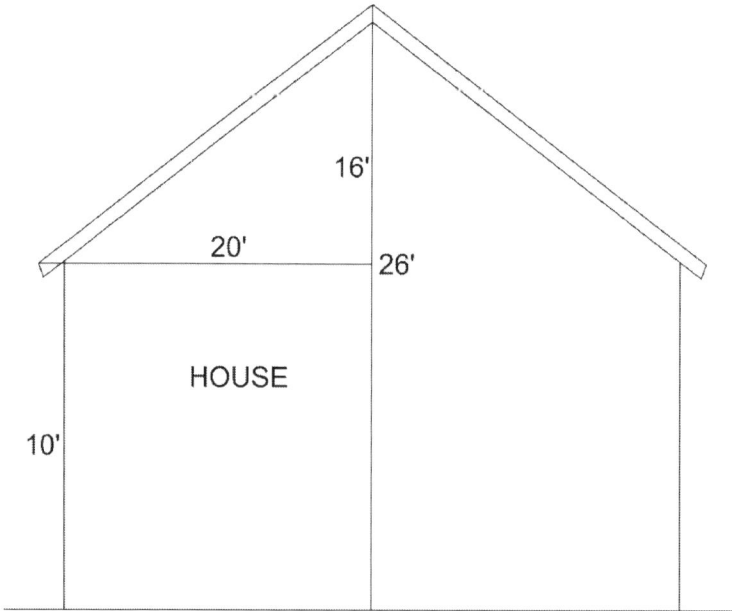

Figure 9 - House Roof Example

We can calculate the length of the roof slope using the Pythagorean theorem. The base of the triangle is 20' (side a) and the height of the triangle is 16' (side b).

So, we have a and b, and we want to solve for side c. We would rearrange Equation 5 as follows:

$$c = \sqrt{a^2 + b^2} \quad \text{so} \quad c = \sqrt{20^2 + 16^2} = 25.6 \text{ Feet}$$

Oblique Triangles

Oblique triangle solutions can get complex, so we will only cover a few commonly used solutions in this book.

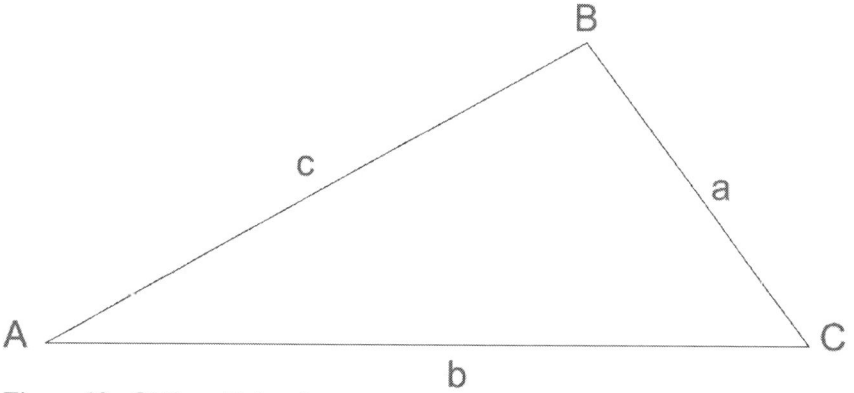

Figure 10 - Oblique Triangle

The solution for oblique triangles is shown in Equation 6. These equations are known as the "Law of Sines".

$$\frac{a}{\sin A} = \frac{b}{\sin B} = \frac{c}{\sin C} \qquad \text{Equation 6}$$

Let us look at the typical solution of the oblique triangle shown in Figure 11. Notice that we are given angles A and C and the length of side a.

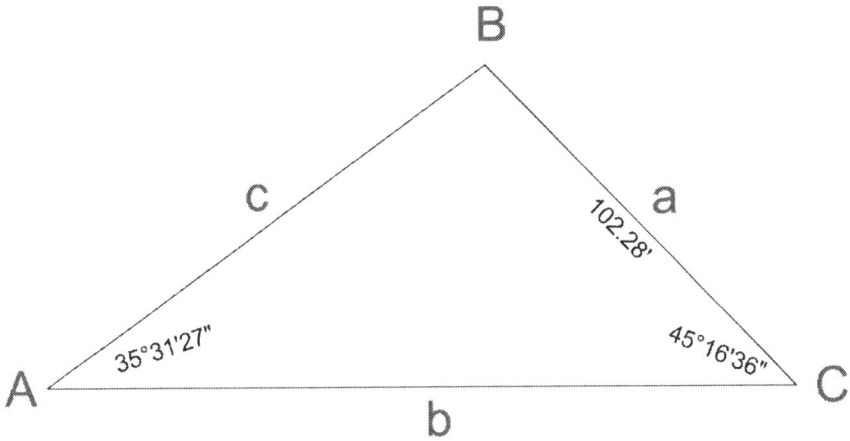

Figure 11 - Oblique Triangle Example

From Equation 6 we see that the following equation can be used:

$$\frac{a}{\sin A} = \frac{c}{\sin C}$$

We can rearrange this equation to solve for unknown variable c.

$$c = \frac{a}{\sin A} * \sin C$$

Substituting our values in the equation we can solve it as follows:

$$c = \frac{102.28}{\sin 35°31'27''} * \sin 45°16'36''$$

$$c = \frac{102.28}{0.5810} * 0.7105 = 125.070$$

We will learn in a later chapter in this book that the sum of the interior angles of any triangle must equal 180°. Notice in Figure 11 that we know two of the angles, so we can calculate the third angle by adding the two known angles together then subtracting the sum from 180° as follows:

33

$$Angle\ B = 180° - (35°31'27 + 45°16'36) = 99°11'57"$$

We can now update our image to show side c and angle B. Now that we know angle B we can solve for the length of the remaining side b.

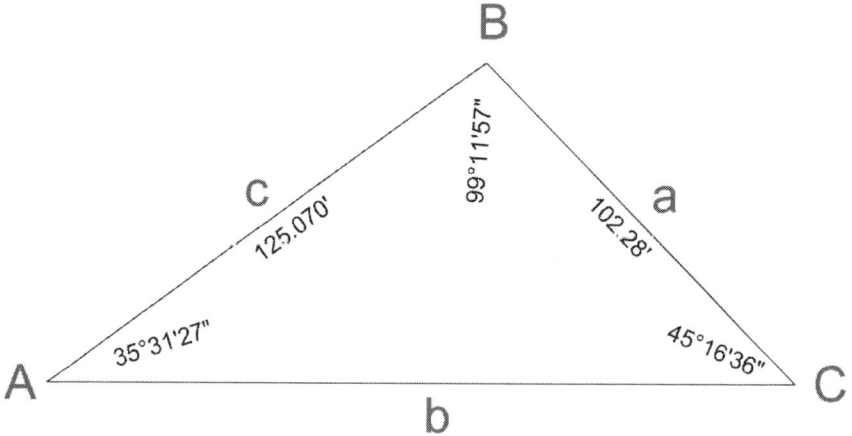

Figure 12 - Oblique Triangle Example 2

Let us solve for side b first. Looking at Equation 6 again, we see that the following equation can be used:

$$\frac{a}{\sin A} = \frac{b}{\sin B}$$

We can rearrange this equation to solve for unknown variable b.

$$b = \frac{a}{\sin A} * \sin B$$

Substituting our values in the equation we can solve it as follows:

$$b = \frac{102.28}{\sin 35°31'27"} * \sin 99°11'57"$$

34

$$b = \frac{102.28}{0.5810} * 0.9871 = 173.763$$

The examples show that some oblique triangles can be solved quite simply using the law of sines and a step by step process. Some solutions however are somewhat more complicated so, for these solutions, the reader is advised to consult a text on trigonometry. We will also learn that instead of solving such triangles using trigonometry, coordinate geometry can also be used in some cases.

Bearings

We stated in *Land Surveying Simplified* that bearings are angles measured from some standard reference direction, such as true north or magnetic north.

When land surveyors use bearings, the circle is divided into four quadrants. Referring to **Figure 13**, we can see the four quadrants, each having a bearing. The term "quadrant' means an arc of 90° which is ¼ of a circle. Bearing directions are always measured from either the north direction or the south direction. So, as shown in **Figure 13**, a bearing of N31°E refers to an angle measured 31° clockwise from north. The bearing N44°W is in the NW quadrant so it is measured counterclockwise from north. The bearing of S79°E is in the southeast quadrant so it is measured counterclockwise from south. The bearing S42°W is in the southwest quadrant so it is measured clockwise from south.

When surveyors write bearings, they are written using Degrees, Minutes and Seconds. An example would be S65° 44' 36"E. One advantage of using this system of quadrants is that it is very easy to change the direction of a bearing by 180°. If you need to reverse the direction of a bearing, we can simply specify the opposite quadrant. For example, a bearing having a direction of N 31° E can be reversed by relabeling it S 31° W.

Many, if not most, plans on record use bearings to fix the direction of boundary lines and traverse lines so surveyors must be proficient at performing bearing calculations. We will look at some of the more common calculations involving bearings and angles.

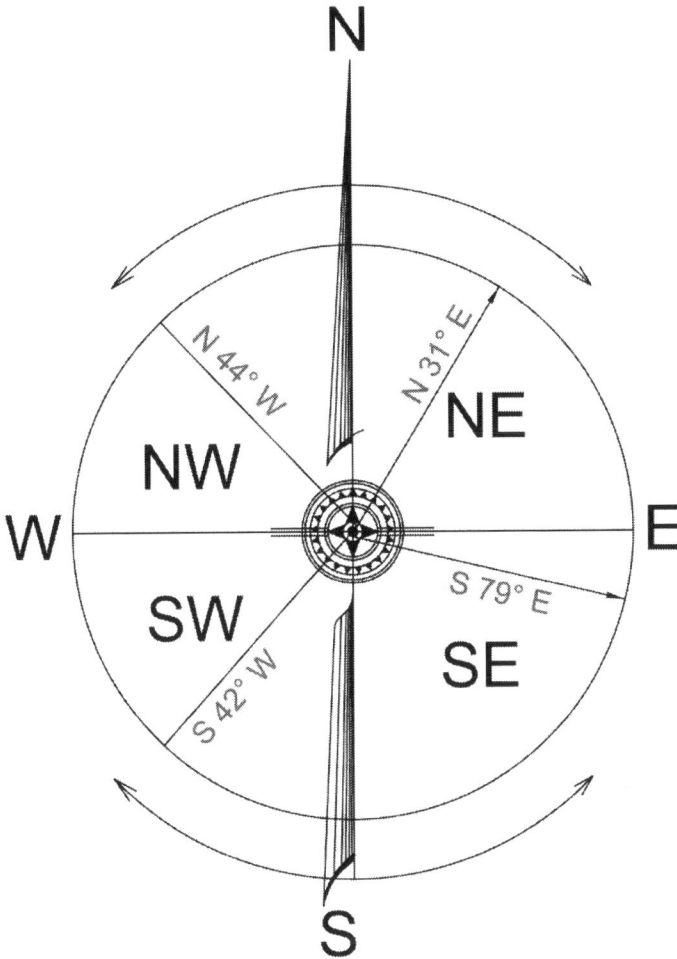

Figure 13 - Bearing Quadrants.

Calculating a New Bearing when we have an Existing Bearing and an Angle.

Situation 1. When the new bearing will be in the same quadrant as the original bearing.

In our first example shown in Figure 14 we have a line fixed in direction by the bearing N75° 55' 37"E.

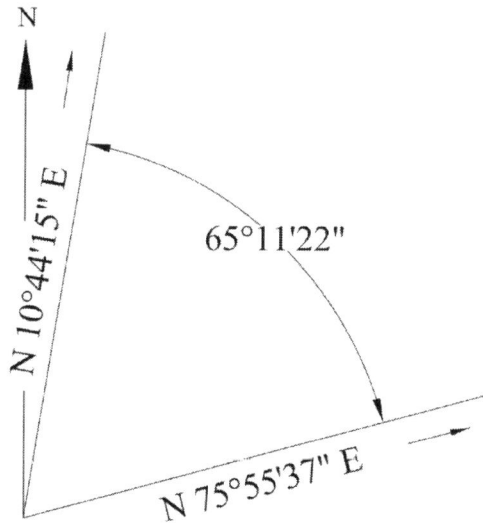

Figure 14 – Calculate a new bearing from a bearing and angle.

Assume that we set up our total station at the vertex shown in the figure at the lower left corner and turn an angle to the left of 65° 11' 22". We want to know the bearing of the new line. The first thing that we notice is that the magnitude of our angle is less than the bearing. In other words, the bearing is 75° and our angle is 65°. This immediately tells us that the new bearing will still be in the NE quadrant. So, to calculate the new bearing we simply subtract the angle from the bearing of the original line as follows:

$$75° \ 55'37" - 65° \ 11' \ 22" = N10° \ 44'15"E$$

For a variation on the above calculation, using the same information shown in Figure 14, assume that the original line was the one labeled N10° 44' 15"E. We again set up our total station at the vertex and this time turn an angle of 65° 11' 22" to the right. We would calculate the bearing of the new line by adding the angle to the original bearing as follows:

$$10° \, 44'15" + 65° \, 11' \, 22" = N75°55'37"E$$

Keep in mind that if the angle measured by the total station were large enough such that the new bearing were greater than 90° our new angle would no longer be in the NE quadrant.

Situation 2. When the original bearing is in a north quadrant and the new bearing will be in the adjacent north quadrant. Or, When the original bearing is in a south quadrant and the new bearing will be in the adjacent south quadrant

The next example is shown in **Figure 15**. In this case, both bearings are in the north quadrants. Our original bearing is N47° 27' 06"W. Let us again assume that we set our total station on the vertex at the bottom of the image and turn an angle to the right of 70°11'22". We notice that the magnitude of angle that we measured is greater than the magnitude of the bearing. In this case we subtract the bearing from the angle in order to calculate the new bearing.

$$70° \, 11'22" - 47° \, 27'06" = N22° \, 44'16" \, E$$

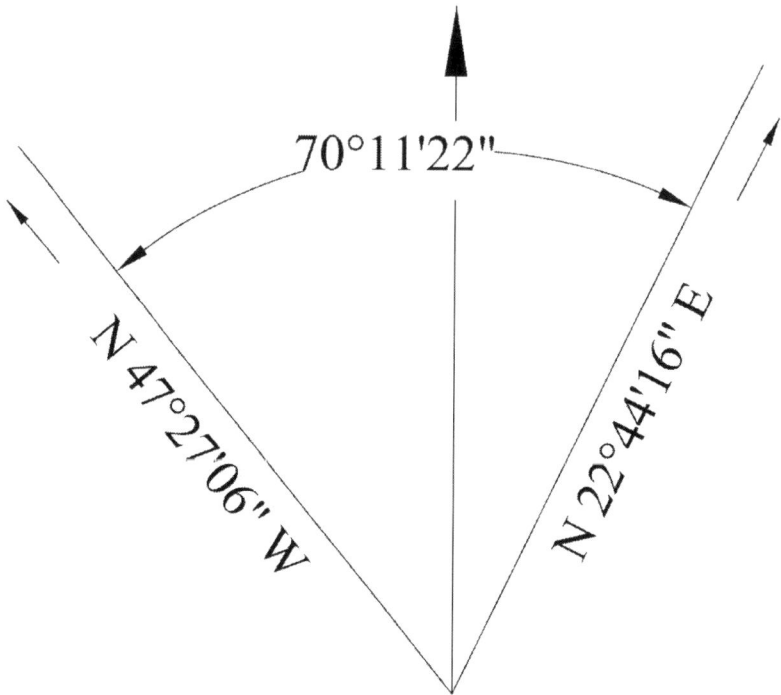

Figure 15 - Calculate a new bearing from a bearing and angle.

Here again, if the resulting bearing were greater than 90° we would not end up in the NE quadrant.

Situation 3. When the new bearing will be in an adjacent quadrant, but the new bearing will be in the opposite half circle, i.e. if the original bearing is in a north quadrant the new bearing will be in a south quadrant or vice versa.

The next situation is illustrated in **Figure 16**. In this case we start with a bearing of N80° 44' 44"E and add an angle to the right of 66° 22' 10". We know that the bearing of N80° 44' 44"E is simply an angle measured clockwise from north. It is also apparent that adding 66° to this angle will place us in the SE quadrant. If we add the bearing and angle together the result will be an angle measured clockwise from north. As long as the resulting angle is

40

less than 180° we can subtract the angle from 180° to get the new bearing. We proceed as follows:

$$80° \, 44' \, 44"" + 66° \, 22' \, 10" = 147° \, 06'54"$$

$$180° - 147° \, 06'54" = S32° \, 53' \, 06"E$$

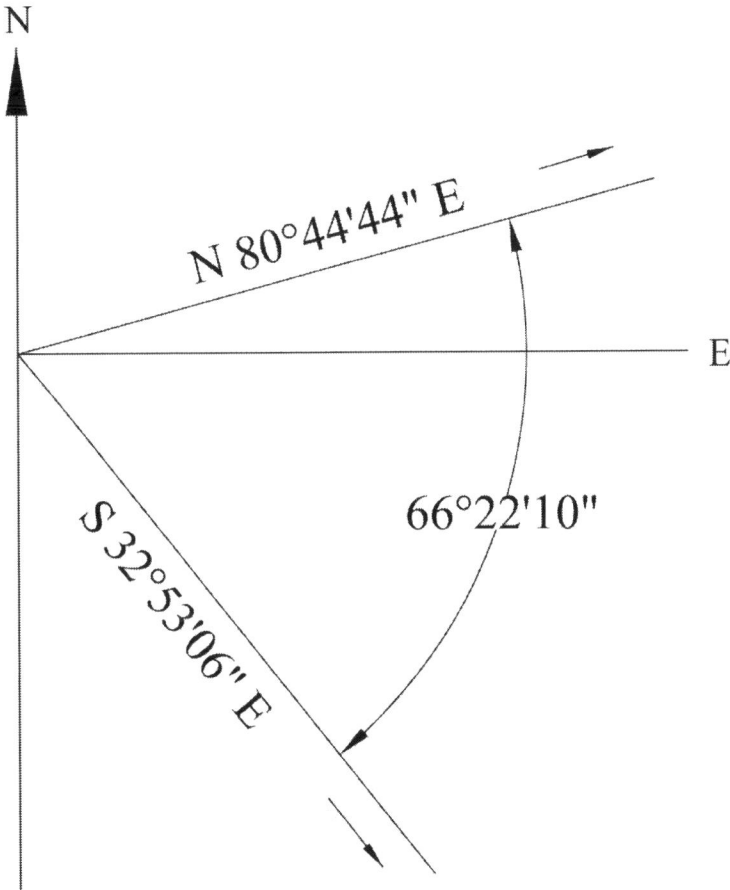

Figure 16 - Calculate a new bearing from a bearing and angle.

41

Situation 4. When the new bearing will be in an opposite quadrant.

Sometimes the new bearing ends up in an opposite quadrant. Consider Figure 17 which shows our initial bearing of S50°36'09"E. We set up our total station on the vertex and turn an angle to right of 159°02'59". An easy way to approach the solution is to realize that we can simply extend the bearing in the SE quadrant to a bearing in the NW quadrant. We already know that the numbers do not change, only the quadrant designation changes.

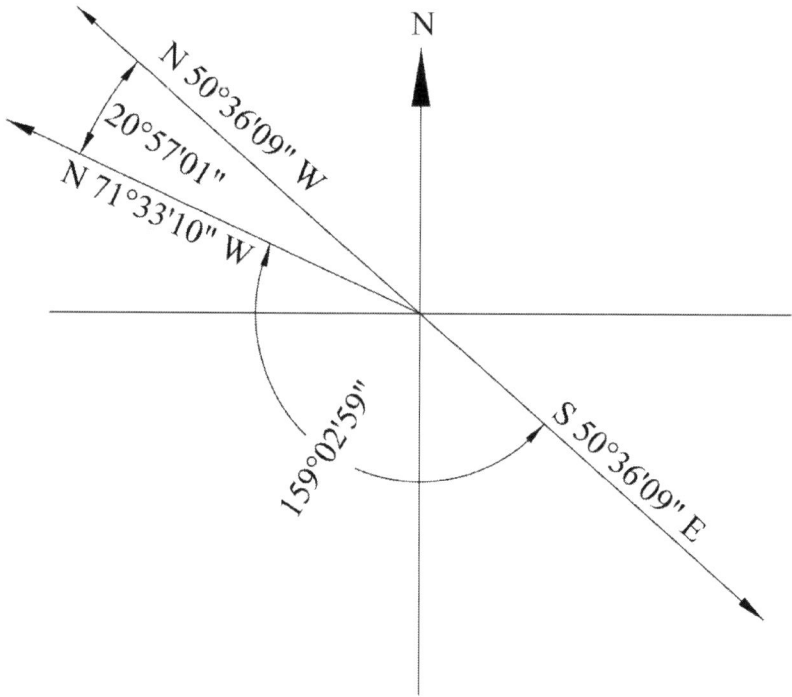

Figure 17 - Calculate a new bearing from a bearing and angle.

You can see from the image that if we subtract our measured angle from 180°, the result will be a deflection angle (described in the next section) from our NW bearing. If we add this angle to the bearing, we will have our new bearing. Here is the calculation:

42

$$180° - 159° 02' 59 = 20° 57' 01$$

$$N50° 36'09"W + 20° 57' 01" = N71° 33' 10"W$$

Calculating an Angle from Two Bearings

The situation opposite to the ones discussed above occurs when we have two bearings and we need to calculate the angle between them. We will use the same examples that were used in the previous section so that you can compare each of the methods.

Situation 1. When the two bearings are in the same quadrant.

In the first situation shown in **Figure 18** both bearings are in the NE quadrant: We calculate the angle by subtracting the smaller bearing from the larger bearing.

$$75° 53'37" - 10° 44' 15" = 65° 11' 22"$$

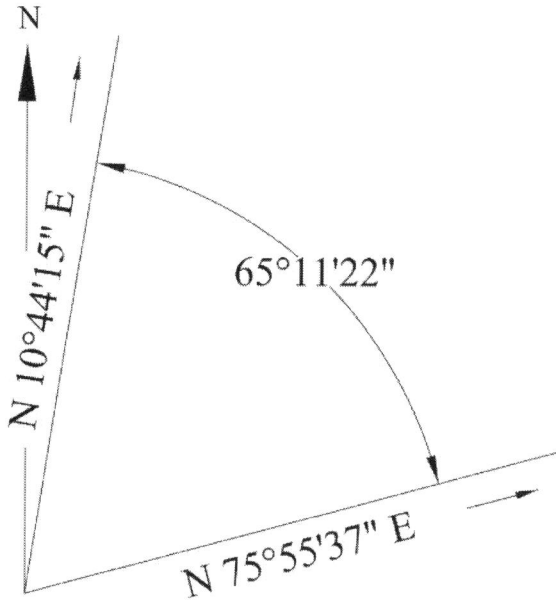

Figure 18.

Situation 2. When the two bearings are in adjacent north or south quadrants.

The next example is shown in **Figure 19**. In this case the bearings are in adjacent north quadrants. Here we simply add both bearings together to calculate the angle. We would use the same method if both bearings were in adjacent south quadrants.

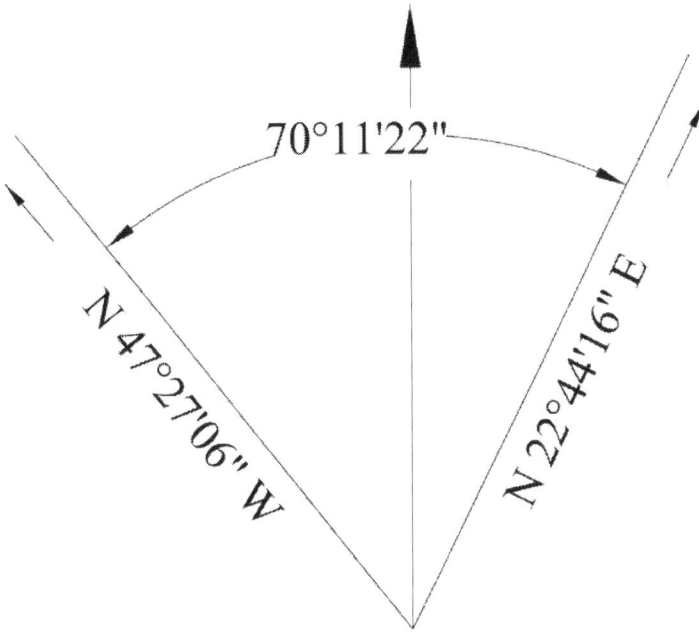

Figure 19.

$$22° 44' 16" + 47° 27' 06 = 70° 11 22"$$

Situation 3. When the two bearings are in adjacent quadrants but one bearing is in a north quadrant and one bearing is in a south quadrant.

The next example is shown in **Figure 20**. Here we have two bearings in adjacent quadrants, but one is in the north quadrant and the other is in the south quadrant. In this case we need to add both bearings together and subtract the sum from 180°.

$$80° 44' 44" + 32° 53' 06 = 113° 37' 50"$$

$$180° - 113° 37' 50" = 66° 22' 10"$$

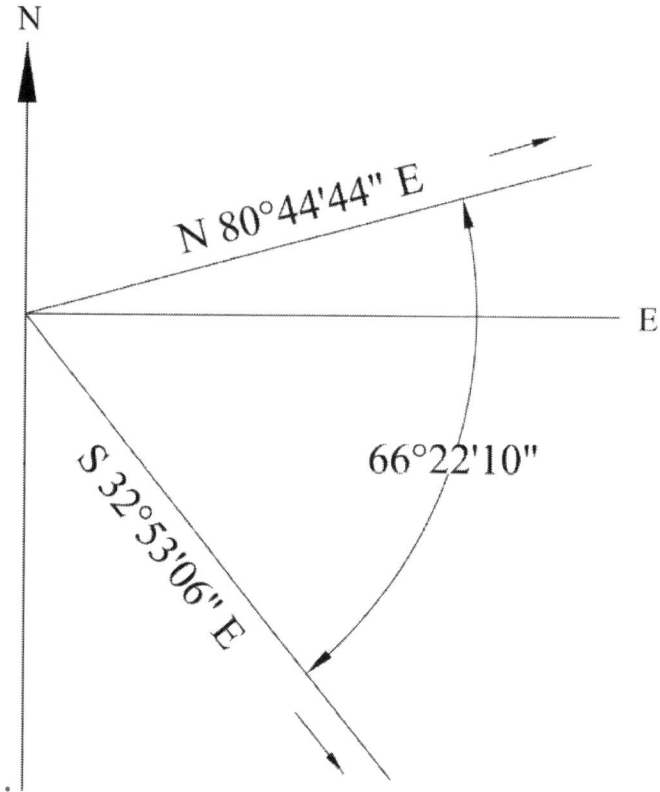

Figure 20.

Situation 4. When the two bearings are in opposite quadrants.

The final example uses two bearings from opposite quadrants as shown in **Figure 21**. As we did in the example shown in Figure 17 we can easily solve this by reversing the direction of one of the bearings so that both bearings are in the same quadrant. We can then subtract one bearing from the other, then subtract the result from 180° to get our angle.

$$71° \, 33' \, 10" \, - \, 50° \, 36' \, 09" \, = \, 20° \, 57' \, 01"$$

$$180° \, - \, 20° \, 57' \, 01" \, = \, 159° \, 02' \, 59"$$

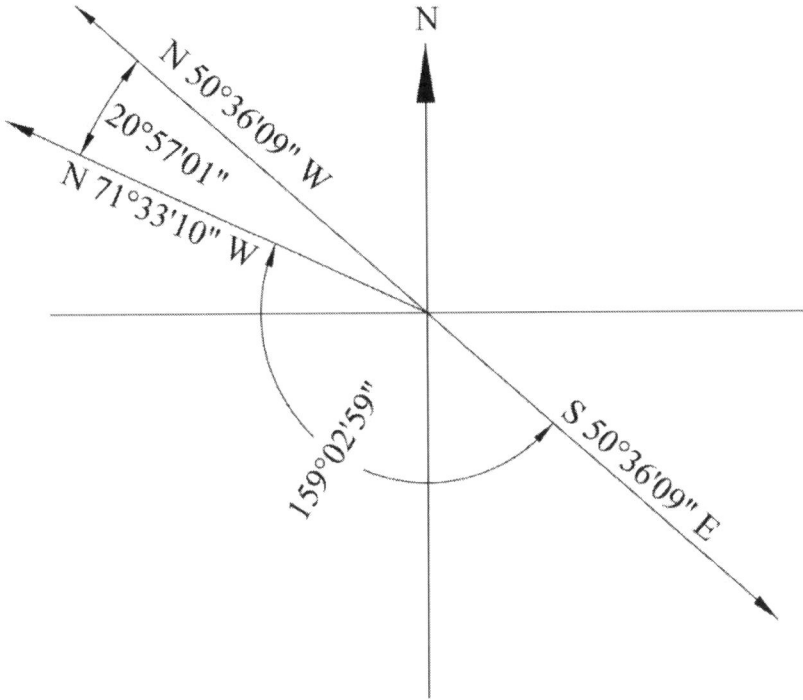

Figure 21.

Deflection Angles.

Deflection angles are angles measured from an extension of the previous line.

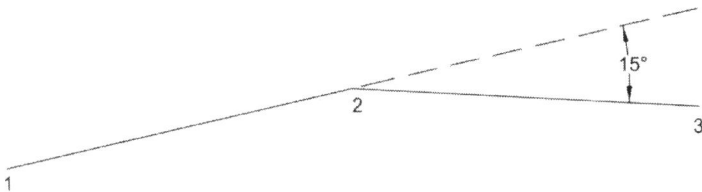

Figure 22 – Deflection angle to the right. *Deflection Angle Right.jpg*

47

Deflection angles may be measured to the right or to the left. **Figure 22** shows a deflection angle to the right.

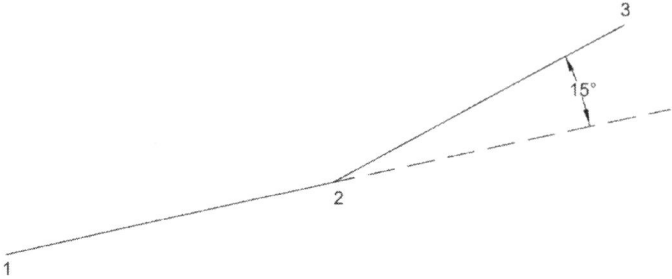

Figure 23 – Deflection angle to the left.

Figure 23 shows a deflection angle to the left. In both of the preceding images the deflection angles deflect from an extension of line 1 to 2 and the deflection angle is 15°.

Angular Closure of a Parcel or Traverse

Although less common today, many old deeds and plans describe line directions using angles not bearings. When a closed traverse is used during the survey of a parcel of land the traverse lines will be fixed by the angles turned by a total station. In such cases we need to be able to determine if the angles close mathematically.

For any closed two dimensional geometric figure, the sum of all of the interior angles must be equal to a certain number. This number will be an integer equal to the number of angles in the figure minus 2 multiplied by 180. Stated in the form of an equation:

$$Sum\ of\ Angles = (n - 2) * 180° \qquad \text{Equation 7}$$

Where n is equal to the number of angles. It is important to realize that this equation only applies to interior angles. If some of the angles happen to be exterior angles then we will have to calculate the interior angle by subtracting the exterior angle from 360°.

Example of an Angular Closure of a Land Parcel.

As an example, consider the plan of land on County Road shown in **Figure 24**.

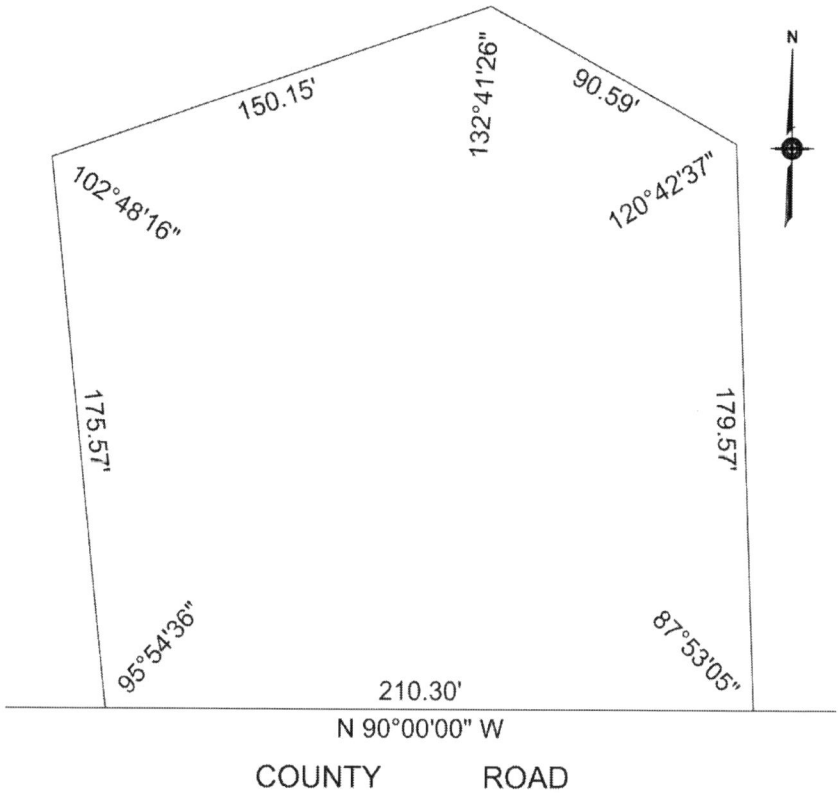

Figure 24 – Parcel defined by angles.

The plan shows a parcel of land described by angles and distances. Because every boundary line intersection is described by interior angles we can check the angular closure of this parcel.

The parcel contains 5 angles. We can use Equation 6 to check the angular closure as follows:

$$Sum\ of\ Angles = (5 - 2) * 180° = 540°$$

We now know that the sum of all of the interior angles in this parcel must equal 540° in order for the parcel to close mathematically. Actually, a mathematical closure will also require that the distances be correct but for now we will just consider the angles. Let us add the angles beginning at the southwest corner of the parcel. Notice that we have displayed the results of the addition of each column.

$$95° \ 54' \ 36"$$
$$102° \ 48' \ 16"$$
$$132° \ 41' \ 26"$$
$$120° \ 42' \ 37"$$
$$\underline{87° \ 53' \ 05"}$$
$$536° \ 238' \ 120"$$

We leave it to the reader to do the math to confirm that we arrive at an answer of:

$$540° \ 00' \ 00"$$

This tells us that the angular closure of the parcel is perfect.

Example of an Angular Closure of a Traverse.

We know that surveyors commonly use a closed traverse when surveying land, particularly when the parcel of land is relatively large or when it is necessary to traverse a city block. A traverse will have at least three interior angles and probably many more, depending on the size of the traverse and on the necessity of avoiding obstructions to the line of sight of the total station.

Even the most carefully made angular measurements will contain small errors. These errors can come from tripods not being set exactly over a point, from small errors in sighting a prism, from atmospheric effects such as heat waves and from countless other effects. In most cases the errors will be relatively small – often just a few seconds of arc. In order to adjust the traverse, these errors must be removed so that the sum of the interior angles of the traverse add up to the correct number. Consider the traverse illustrated in **Figure 25**. The angles shown in this traverse consist of the raw data measured by the total station.

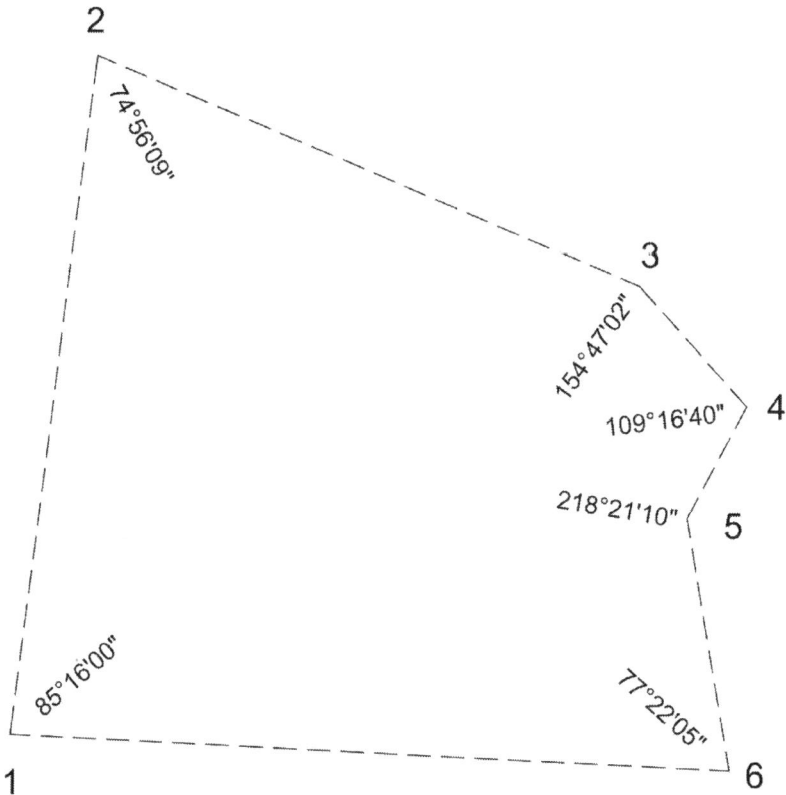

Figure 25 – Traverse Angular Closure.

The traverse contains 6 angles, so the sum of the angles should be:

$$Sum\ of\ Angles = (6 - 2) * 180° = 720°$$

The first step will be to add together all the interior angles to see how closely the actual measurements agree with the theoretical value.

Point	Angle
1	85° 16' 00"
2	74° 56' 09"
3	154° 47' 02"
4	109° 16' 40"
5	218° 21' 10"
6	77° 22' 05"
Sum	719° 59' 06"

Figure 26 –Angles as measured in the field.

Instead of 720° the sum is 719° 59' 06". Our actual angular measurement sum is in error by 54 seconds.

$$720° \, 00'00" - 719° \, 59' \, 06" = 00° \, 00'54"$$

A simple way to adjust the angles would be to divide the error by the number of angles and apply the adjustment equally to each angle. Although this is a commonly used procedure, particularly when the error is small, it probably will not place the errors where they truly exist. For example, if you were the person who measured the angles in the field, you may be aware of a particular instrument setup that was on unstable ground. You may have turned multiple angle sets and know that one or more angles did not agree very well. The point is that if there is reason to suspect an error in a particular angle it may be prudent to place the closure error in that angle. Alternatively, if the error is large enough it may be necessary to make a return trip to the field in order to remeasure one or more angles. For our purposes here, we will simply apply the error equally to all angles.

Our raw angular data shows that the sum of the 6 angles is 54 seconds smaller than the correct value. Dividing 54 seconds by 6 means that we must add 9 seconds to each angle. The results are shown in **Figure 27**.

Point	Angle	Adjustment	Adjusted Angles
1	85° 16' 00"	09"	85° 16' 09"
2	74° 56' 09"	09"	74° 56' 18"
3	154° 47' 02"	09"	154° 47' 11"
4	109° 16' 40"	09"	109° 16' 49"
5	218° 21' 10"	09"	218° 21' 19"
6	77° 22' 05"	09"	77° 22' 14"
Sum	719° 59' 06"		720° 00' 00"

Figure 27 – Adjusted angles.

Once we have adjusted our angles we can proceed to adjust the traverse. In some cases, if the angular error is very small, some surveyors will skip the angle adjustment procedure and simply adjust the traverse. We will see that this will adjust both the angles and the distances.

Coordinate Geometry

All land surveyors must have a complete mastery of coordinate geometry. Coordinate geometry forms the basis of nearly all calculations involving boundaries and other features which surveyors must work with.

Surveyors use coordinate geometry to check the closure of deed descriptions. Coordinate geometry is also used by surveyor to close and to adjust traverses. It is used to plot the locations of physical evidence found in the field and to make determinations of how well physical evidence fits record evidence. It is used to design divisions and subdivisions of property. It is used to calculate road layouts. Coordinate geometry is the primary calculation tool used by surveyors on a daily basis.

We learned about data collection in *Land Surveying Simplified*. Data collection performed in the field using a total station records the angles and distances to traverse points and to physical evidence or other objects that a surveyor wishes to locate. If the only calculation tool available to a surveyor were the angles and distances recorded in the field it would exceedingly difficult to mathematically determine the relationships between the physical evidence and the record bearings and distances which describe the property being surveyed. Coordinate geometry allows surveyors to use the angles and distances recorded in the field to calculate "coordinates" for every point which was located in the field. Each point will have its own unique coordinate. These coordinates will allow a surveyor to easily and precisely determine the relationship between points. This chapter will explain the principles underlying coordinate geometry and how surveyors use coordinate geometry to perform these calculations.

We have learned that when a total station is used to locate a point it measures and records three pieces of information: The angle to the point, the distance to the point and the vertical angle to the point. Recall that because total stations measure slope distances, knowing the vertical angle is necessary in order to calculate the horizontal distance to the point. For the purposes of most boundary surveys which do not require elevation information, this is the primary use of the vertical angle. Nearly all boundary dimensions are two dimensional. A notable exception is when a boundary is tied to some elevation such as the boundary along a shoreline which is based on mean high water or mean low water. But even this information is usually shown on plans in two dimensions. Because of the two dimensional nature of boundary surveying and because working in two dimensions rather than three simplifies the explanation of coordinate geometry we will limit our discussion to two dimensional systems.

The Meridian of the Survey

Coordinate geometry used by land surveyors requires that the survey be related to some meridian, usually north. It could be magnetic north, true north, grid north or some rough approximation of north. For internal calculation purposes, the precision with which north is determined is largely irrelevant because, as we will see, what we are most interested in at this stage is the relationship between points – and this is independent of precisely which way north points. Once we have determined these relationships we can rotate our entire survey to some absolute meridian if this becomes necessary.

Cartesian Coordinates

Surveyors use Cartesian Coordinates. The Cartesian Coordinate system is a two-dimensional coordinate system having two axes 90° apart. The system is essentially a rectangular grid such as the one shown in Figure 28. The axes are labeled **X** for the horizontal axis and **Y** for the vertical axis. Traditionally, the Y axis represents the north-south axis and runs vertically up the sheet.

The X axis represents the east-west axis and runs horizontally across the sheet. Values increase up and to the right.

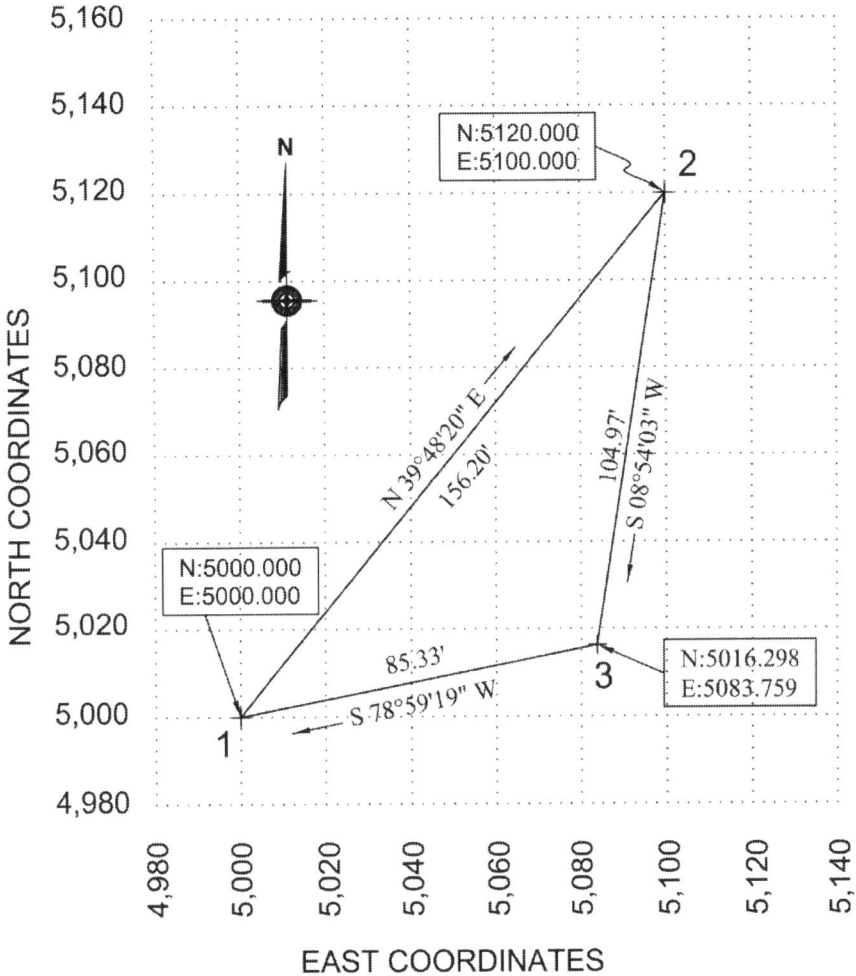

Figure 28 – Coordinates of a line.

In land surveying, the Y axis is labeled the **Latitude (Northing)** and the X axis the **Departure (Easting)**. Referring to Figure 28, Latitudes increase in a northerly direction. Departures increase in an easterly direction. The term **Coordinate** refers to the pair of numbers that fix the location of a point on the grid. It is the Latitude and Departure of the point. In surveying calculations, the coordinates of a point are labeled N for northing and E for easting.

An important thing to understand about the Cartesian Coordinate system is that a point is precisely fixed in position by its two-dimensional coordinate. The two-dimensional coordinate is an intersection of the northing value (latitude) with the easting value (departure). You can see from the figure that point number 1 has a coordinate of N5,000.000, E5000.000. Clearly there is only one location on the grid that a point with these coordinates can exist. It is at the intersection of the 5,000 grid line running north-south and the intersection of the 5,000 grid line running east-west.

Choosing a Beginning Coordinate

Let us assume that the triangle shown in Figure 28 is a triangular parcel of land described in the following deed description:

"...Beginning at a point at the southwest corner of the land to be conveyed thence running N39° 48' 20"E, a distance of 156.20 feet; thence running S08° 54' 03"W a distance of 104.97 feet; thence running S78° 59' 19"W 85.33 feet to the point of beginning..."

The deed describes the three consecutive boundaries using bearings and distances. These boundaries are shown in Figure 28 beginning with point number 1 and proceeding to point 2, then to 3 and continuing to the point of beginning.

Deeds rarely contain coordinate values for boundary corners and you can see from the deed description that the coordinates in our figure do not arise from the deed. The beginning coordinate value that we have used is completely arbitrary. The beginning coordinate value that we use does not matter because we will primarily be interested in the differences between the coordinates not the absolute value of the coordinates. The only real issue when choosing which coordinate value to start with is we want to pick a value that is large enough so that we do not end up with negative coordinate values. There is nothing inherently wrong with negative coordinate values other than it is a nuisance having to deal with minus signs when there is no valid reason for doing so. In our example, when we read our deed, we immediately realize that we are dealing with a small parcel of land less than 200 feet in size so by picking 5,000/5,000 as a starting coordinate value we are assured that all of the coordinates describing the parcel will be positive. We could have chosen 1,000/1,000 or even 500/500. Experienced surveyors usually choose a relatively large value because they realize it may be necessary to go back into the field to locate a monument which might be 1,000 feet away. If the monument is to the west or south of the locus, the point could end up with a negative value if a small starting coordinate value was used.

We have already discussed the importance of right triangles in surveying. Right triangles are an essential part of coordinate geometry. We will use a right triangle to calculate the latitude and departure of our first line. This will enable us to calculate coordinates for point 2.

Calculating the Coordinates of Point 2

In our example, we have chosen N5,000.000, E5,000.000 as our beginning coordinate for Point 1. Our next step is to calculate the coordinates of the remaining two corners. We have already covered right triangles in an earlier section. It turns out that our knowledge of right triangles will allow us to calculate the coordinates of the remaining two corners. In doing so, it will be helpful to characterize the line as the hypotenuse of a right triangle as shown in Figure 29. Instead of the standard representation of a right triangle shown in Figure 5, we have rotated it so that the hypotenuse is oriented in the same direction as the first boundary described in the deed for our parcel of land (N39°48'20"E). Instead of the base of the triangle (b) being at the bottom it is now on the right.

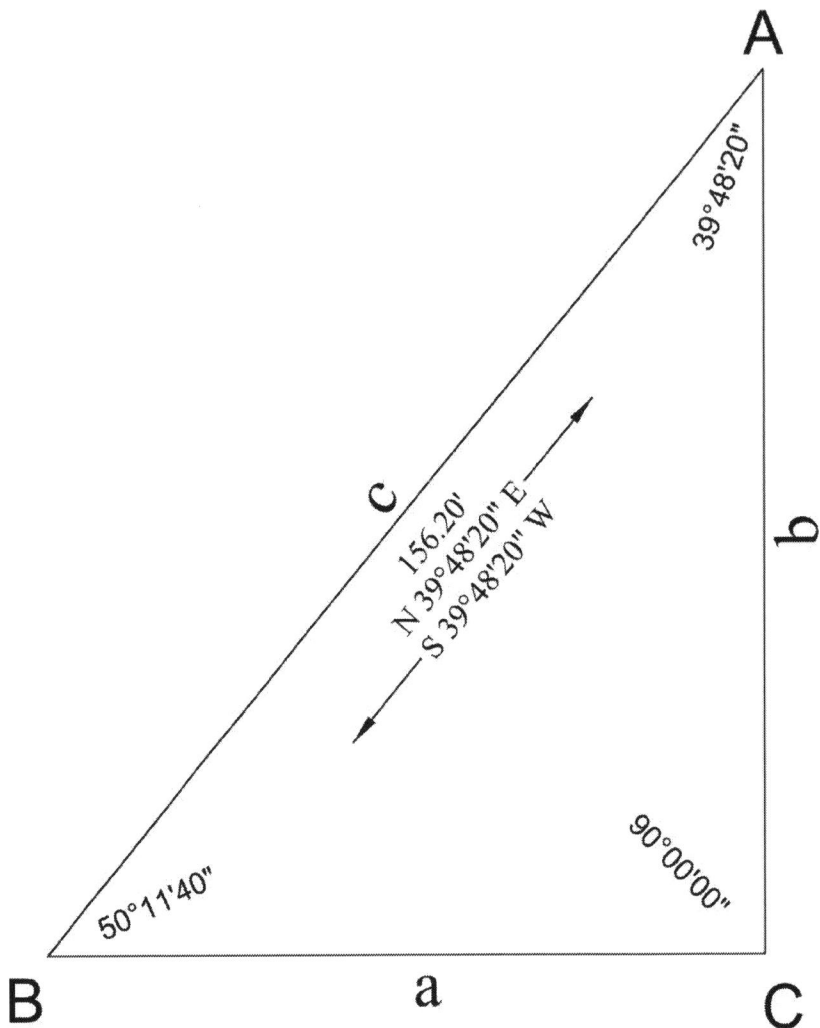

Figure 29 – Using a right triangle to calculate latitude and departure.

We have oriented our right triangle so that side b runs north-south and side a runs east-west. The direction of these two sides corresponds exactly with our Cartesian coordinate grid lines. We can use the bearing of our first boundary line (N39°48'20"E) to calculate the interior angles of the triangle. Recall that the direction of a bearing in the NE quadrant can be reversed to the SE quadrant simply by changing the quadrant designation, in this case to S39°48'20"W. In the section on bearings we learned that the bearing of a line in the southwest quadrant is simply an angle measured clockwise from due south. This tells us that angle A in our right triangle is the same value as our bearing: 39°48'20".

If we need to know the value of angle B, we can calculate it because we know from Equation 7 that, for any closed geometric figure, the number of interior angles minus 2 times 180° equals the sum of the angles. The sum of the interior angles of a triangle must, by definition, equal 180° ([3 angles -2] *180 = 180). In our example we know angle A and angle C, so we calculate:

$$Angle\ B = 180° - (39°48'20" + 90°) = 50°11'40"$$

We start at Point 1 in Figure 28 (the location of Angle B in Figure 29) and use the bearing and distance of N39°48'20"E, 156.20' to calculate the coordinate of point 2.

Let's solve the right triangle for side "a" first. We know angle A and side c and our unknown is side a. This means that we will want to use **Equation 1** because it contains these three variables:

$$sin\ A = \frac{a}{c}$$

We need to solve for side a so we must rearrange the equation:

$$a = c * sin\ A$$

Substituting the values:

$$a = 156.20' * sin(39°48'20") = 100.000'$$

Now, let's solve for side "b". We know angle A and side c and our unknown variable is side b. Equation 2 contains these three variables.

61

$$\cos A = \frac{b}{c}$$

We need to solve for side b so we must rearrange the equation:

$$b = c * \cos A$$

Substituting the values:

$$b = 156.20' * \cos(39°48'20") = 120.000'$$

The two values that we have just calculated are the values that must be applied to the coordinates of Point 1 in order to obtain the coordinates of Point 2. These are the latitudes and departures that we discussed earlier. In our example the latitude (northing) is 120.00' and the departure (easting) is 100.00'. We can see from Figure 28 that Point 2 is north of Point 1 and we know that coordinate values increase in a northerly direction, so we must add the latitude to Point 1. Similarly, we know from the figure that Point 2 is east of Point 1, and coordinate values increase in an easterly direction, so we must add the departure to Point 1. The math is as follows:

Point 2 N coordinate $= 5,000.000 + 120.000' = 5,120.000$

Point 2 E coordinate $= 5,000.000 + 100.000' = 5,100.000$

Calculating the Coordinates of Point 3

We will use the same procedure just described to calculate the coordinates of Point 3. The right triangle is shown in Figure 30. The bearing of the boundary line is angle A (8°54'03").

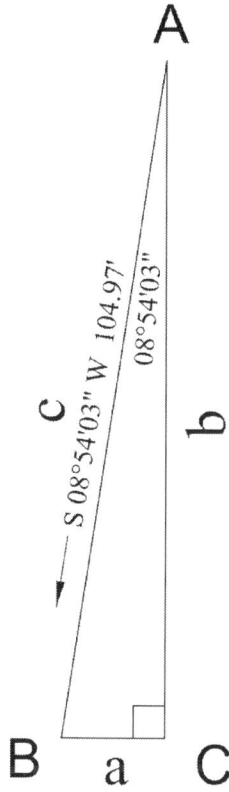

A

c
S 08°54'03" W 104.97'
08°54'03"

B   a   C

b

Figure 30

We use our first equation to solve for side a:

$$a = c * \sin A$$

Substituting the values:

$$a = 104.97' * \sin(8°54'03") = 16.241'$$

Next, solve for side "b".

$$b = c * \cos A$$

63

Substituting the values:

$$b = 104.97' * cos(8°54'03") = 103.706'$$

Next, calculate the coordinate values by applying the latitude and departure to Point 2. Notice that this boundary line runs southwest from Point 2 so we must subtract the latitude and departure from Point 2 coordinates.

Point 3 N coordinate = 5,120.000 − 103.706' = 5,016.298

Point 3 E coordinate = 5,100.000 − 16.241' = 5,083.759

If you perform these same calculations you may find that there are small differences due to rounding errors.

Calculating Coordinates using a Tabular Format

Although the procedure that we used above to calculate coordinates has educational value and gets the job done, creating right triangles for each point is a laborious and time-consuming method. Still, the lesson does bring to light some important points. First, latitudes are calculated using the cosine of the angle and departures are calculated using the sin of the angle. Second, the direction in which a line is running will determine whether we need to add or subtract a latitude or departure. Third, because we are proceeding around a parcel boundary or traverse in a sequential manner, latitudes and departures for a new point are always applied to the previous point. We can use this knowledge to create a table in which to perform our calculations. This will standardize our procedure to make our job simpler and easier and help to eliminate errors, such as inadvertently adding rather than subtracting a latitude or departure.

Looking again at Figure 29, we see that side "b" of our right triangle is the latitude and side "a" is the departure. We can use this information to write the following equations:

$$Latitude = Distance * cos(Bearing) \qquad \text{Equation 8}$$

$$Departure = Distance * sin(Bearing) \qquad \text{Equation 9}$$

Using the line shown on **Figure 28**, we can create a table as in **Figure 31**. In this table, you can see how the start coordinates, the latitude and departure and the end coordinates are calculated and arranged. Using such a table like this is much simpler and more convenient than constructing right triangles for each course.

COGO 1 Traverse Table

Point	Bearing	Distance	Latitude (cos)		Departure (sin)		Coordinates	
			North +	South -	East +	West -	N	E
1							5,000.000	5,000.000
	N39° 48' 20"E	156.200	120.000		100.000			
2							5,120.000	5,100.000

Figure 31 – Latitudes and Departures used to calculate coordinates.

Notice that the point number is on its own line along with the coordinates for that point. Also notice that the latitudes and departures are labeled + and – so that we will know whether to add or subtract them from the coordinate of the previous point. If, as in our example, the line runs North-East, both values will be positive and both the North and East coordinates of the point being calculated will increase. If our line ran South-East, the north coordinate would decrease but the East coordinate would increase.

Let us take a look at a slightly more complex example. The example shown in Figure 32 is a parcel of land having 5 boundaries.

65

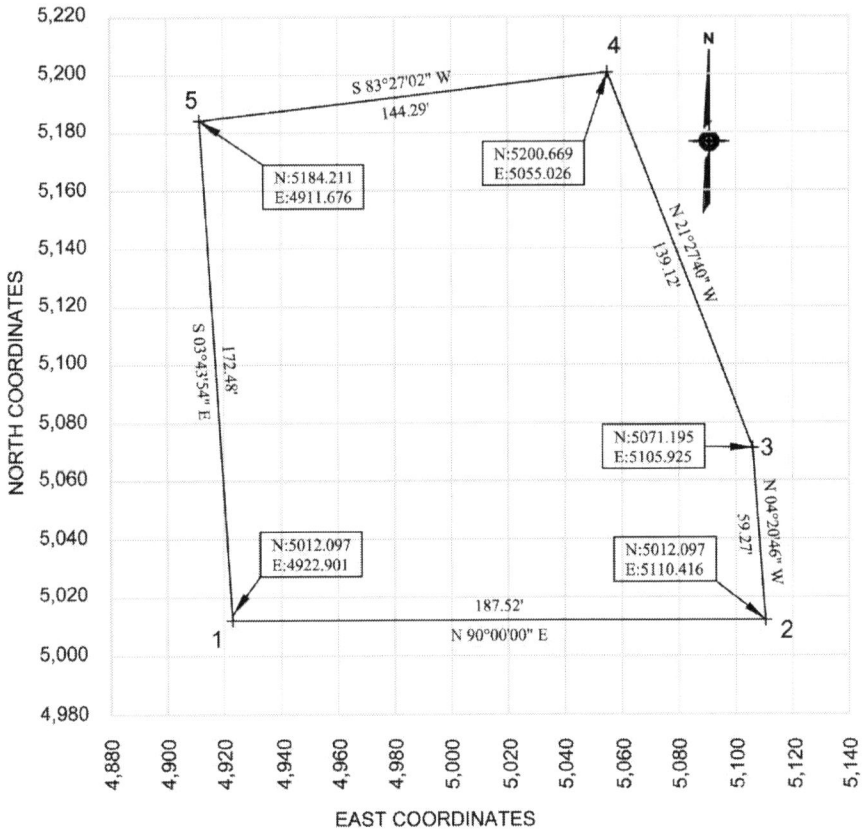

Figure 32 – A parcel of land with four boundaries.

The coordinate table is shown in **Figure 33.** Notice that north latitudes increase the values of the north coordinates and south latitudes decrease their values. Notice that east departures increase the values of east coordinates and west values decrease their values. Also notice that the sums of the north and south latitudes are the same, but with opposite signs to they add up to zero. This is because we started at Point 1 and returned exactly to Point 1. The same is true for the sums of the east and west departures. Also notice that the coordinates for Point 1 are exactly the same at the beginning and end. This tells us that the boundaries of this parcel

66

close perfectly. Said in another way, the error of our closure is zero.

COGO 3 Traverse Table

Point	Bearing	Distance	Latitude (cos)		Departure (sin)		Coordinates	
			North +	South -	East +	West -	N	E
1							5,012.097	4,922.901
	N90°00'00"E	187.520	0.000		187.520			
2							5,012.097	5,110.416
	N04°20'46"W	59.270	59.098			-4.491		
3							5,071.195	5,105.925
	N21°27'40"W	139.120	129.474			-50.899		
4							5,200.669	5,055.026
	S83°27'02W	144.290		-16.458		-143.350		
5							5,184.211	4,911.676
	S03°43'54"E	172.480		-172.114	11.225			
1							5,012.097	4,922.901
		Sums	188.572	-188.572	198.745	-198.740		

Figure 33 – Table showing latitudes and departures.

Inversing – Calculating the Bearing and Distance Between Two Points

Sometimes we need to calculate a bearing and distance between two points. Surveyors call this an **Inverse**. In order to inverse between two points, we need to have the coordinates of each point. Let's look at an example of an inverse. Consider the example shown in Figure 34. We want to inverse from Point 1 to Point 2. We know the coordinates of Point 1 and Point 2. The bearing and distance that we need to calculate is shown in the figure as the hypotenuse of the triangle.

67

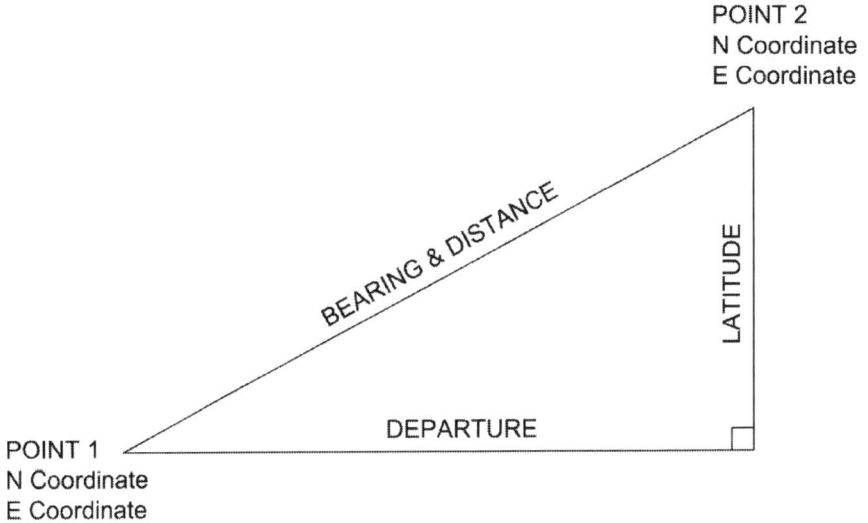

POINT 2
N Coordinate
E Coordinate

BEARING & DISTANCE

LATITUDE

DEPARTURE

POINT 1
N Coordinate
E Coordinate

Figure 34 – Right Triangle used to visualize an inverse.

Our first step will be to calculate the latitude and departure. We can use the following two equations.

$$Latitude = Point\ 2\ N - Point\ 1\ N \qquad \text{Equation 10}$$

$$Departure = Point\ 2\ E - Point\ 1\ E \qquad \text{Equation 11}$$

Where N and E are the North and East coordinates of the point.

Now that we know the latitude and departure we have enough information to calculate the bearing of the line. The following equation will give is the bearing:

$$Bearing = \tan^{-1}\left(\frac{Departure}{Latitude}\right) \qquad \text{Equation 12}$$

Recall from our discussion in the trigonometry section that $\tan^{-1}$ is the arctangent of an angle. Once we have the arctangent we can find the corresponding angle (bearing).

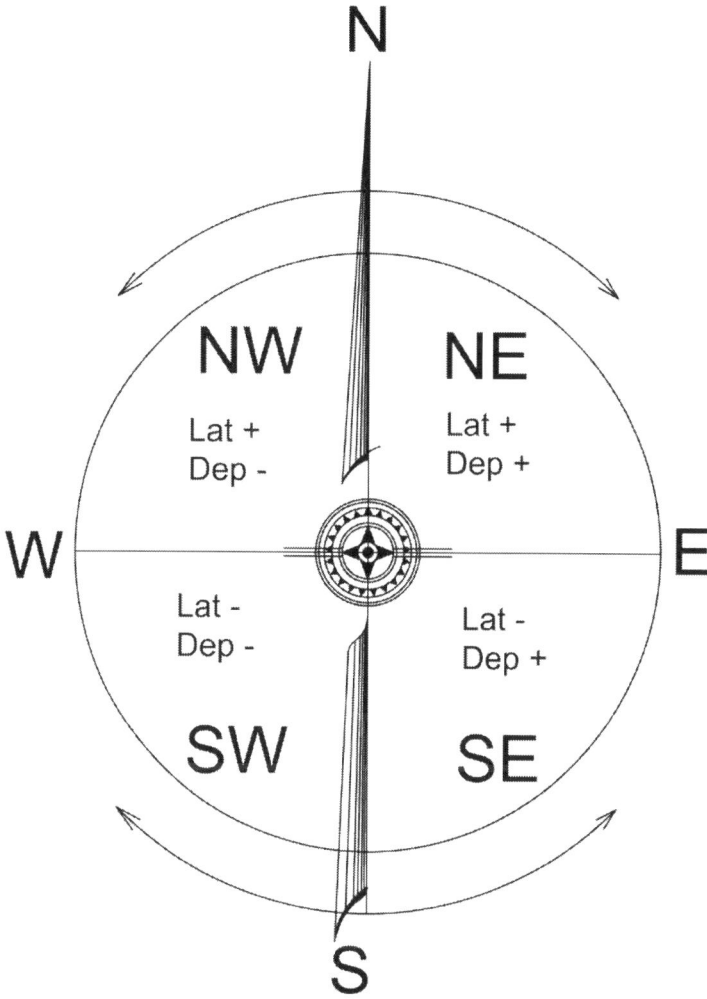

Figure 35 - Latitude & Departure Signs

When doing an inverse, it is important to keep track of positive and negative values of the latitudes and departures so that we will know the direction of the bearing. The image in Figure 35 shows the signs of the latitudes and departures in each quadrant.

Next, we need to calculate the distance. We can use Pythagoras to solve for the distance.

$$Distance = \sqrt{Latitude^2 + Departure^2}$$ Equation 13

Now that we understand the theory of inversing, let us look at an actual example. We will use our earlier example, now shown in Figure 36. Assume that we do not know the bearing and distance of the line running from point 5 to point 4.

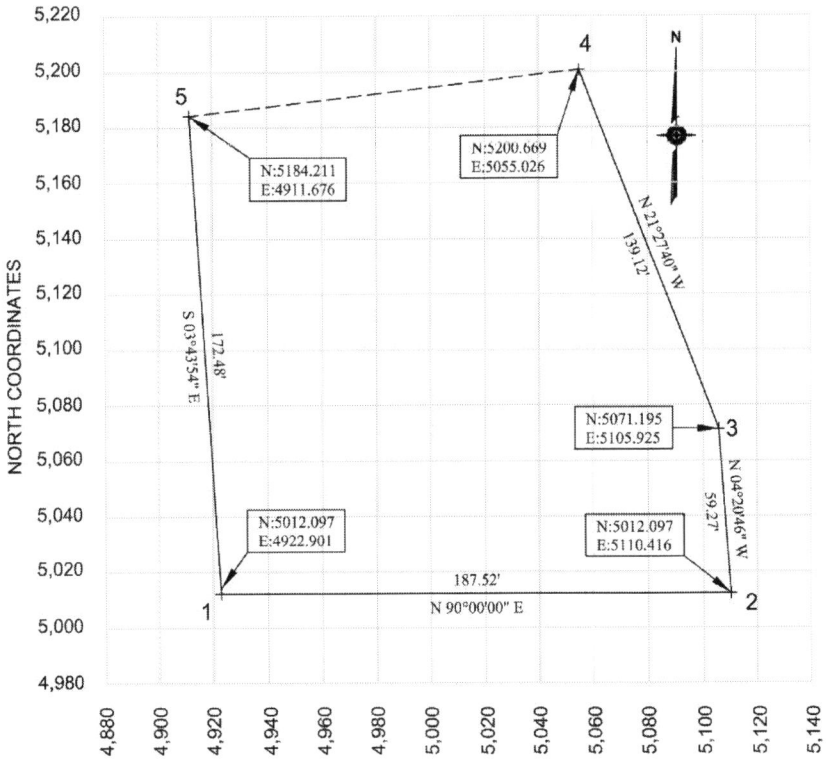

Figure 36 – An Inverse.

Our first step is to set up a table just like the one we used to calculate the coordinates in our previous examples. The table is shown in Figure 37. Although the table is shown with all of the calculations already performed, in actuality at this point we would not know the latitude and departure or the bearing and distance. We would only know the point numbers and coordinates.

Now that we have the coordinates of both points in our table we can calculate the latitude and departure by finding the difference between the coordinates using Equation 10 and Equation 11. You can see from the image that our line direction will be northeast. You can also tell this from the coordinates because both the north and east coordinate increase in magnitude from point 5 to point 4. You can confirm this by looking at Figure 35. When we subtract the north coordinate of point 5 from the north coordinate of point 4 we get a latitude of 16.458'. When we subtract the east coordinate of point 5 from the east coordinate of point 4 we get 143.350'. Both values are positive, so the latitude goes in the North column and the departure goes in the East column.

			Latitude (cos)		Departure (sin)		Coordinates	
Point	Bearing	Distance	North +	South -	East +	West -	N	E
5							5,184.211	4,911.676
	N83°27'02"E	144.290	16.458		143.350			
4							5,200.669	5,055.026

Inverse Table

Figure 37 - Inverse Table

Now that we have the latitude and departure of the line we can calculate the bearing of the line. We use Equation 12 to calculate the bearing as follows:

$$Bearing = \tan^{-1}\left(\frac{Departure}{Latitude}\right)$$

$$Bearing = \tan^{-1}\frac{143.350}{16.458} = 83°\ 27'02"$$

Because our bearing is in the northeast quadrant the bearing is:

$$N83°\ 27'02"E$$

Next, we will calculate the distance. We will use Equation 13 to calculate the distance:

$$Distance = \sqrt{Latitude^2 + Departure^2}$$

Substituting:

$$Distance = \sqrt{16.458^2 + 143.350^2} = 144.290$$

Traverse Adjustment

A number of methods exist for adjusting a traverse. One of the most common methods is the **Compass Rule**. The compass rule assumes that both the angles and distances were measured with similar precision. Other methods for adjusting traverses include the Transit Rule, Crandall Method and Least Squares analysis. In order to keep this book concise, we will only consider the Compass Rule.

Reliance on any of the methods of adjustment may not result in the best possible adjustment if the surveyor is aware of a likely source of error in the traverse. For example, a particular setup may be known to have been made on shaky ground and the angle sets that were measured may have indicated imprecision in the measurement. In such cases it may be necessary to repeat the measurements before attempting to adjust the traverse. Relying on adjustment procedures in such a case would only distribute the error to angles and distances that were not in error to begin with.

Calculating the Error of Closure

Our example traverse is shown in **Figure 38**. This figure contains the raw data which was measured in the field.

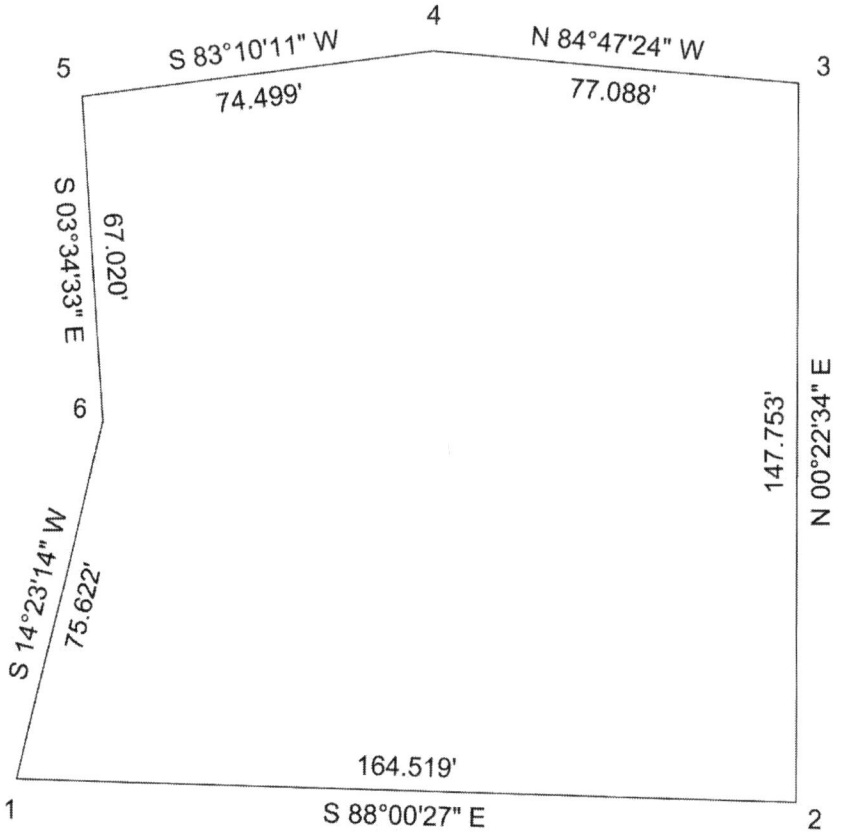

Figure 38 – Raw Data.

The data has been copied to our familiar table format which shows the latitudes and departures and the coordinates. The table is shown in **Figure 39.**

Traverse Adjustment Raw Data Table

Point	Bearing	Distance	Latitude		Departure		Coordinates	
			North +	South -	East +	West -	N	E
1							5000.000	5000.000
	S 88°00'27" E	164.519	-5.720		164.420			
2							4994.280	5164.420
	N 00°22'34" E	147.753	147.750		0.970			
3							5142.030	5165.390
	N 84°47'24" W	77.088	7.000			-76.770		
4							5149.030	5088.620
	S 83°10'11" W	74.499	-8.860			-73.970		
5							5140.170	5014.650
	S 03°34'33" E	67.020		-66.890	4.180			
6							5073.280	5018.830
	S 14°23'14" W	75.622		-73.250	-18.790			
7							5000.030	5000.040
	Sum of Lat and Dep.		140.170	-140.140	146.600	-146.560		
	Difference Lat and Dep.		0.030		0.040			

Figure 39 – Traverse Raw Data Table

The traverse does not close perfectly. The north coordinate is off by 0.03. The east coordinate is off by 0.04'. The last point in the table is numbered Point 7 because the coordinates differ slightly from Point 1 so it is technically a different point. If the traverse had closed perfectly the closing coordinates of Point 7 would be exactly the same as Point 1 (5,000.000, 5,000.000).

The actual error of closure is derived from the differences between the starting and ending coordinates. The bearing and distance of the error can be calculated by inversing between Point 7 and Point 1. The error of closure can be thought of as a line connecting the ending point of the traverse with the beginning point of the traverse. In the example it would be a line connecting Point 7 to Point 1. Because the error is so small in proportion to the scale of the drawing it would not be visible in the image. In our example, the latitude difference is 0.03' and the departure is 0.04'. We are already familiar with how to perform an inverse using **Equation 12**:

$$\tan A = \frac{Departure}{Latitude}$$

Substituting:

$$\tan A = \frac{0.040}{0.030} = 1.333 = 53°\ 07'48"$$

The distance error can be calculated using Equation 13:

$$distance = \sqrt{latitude^2 + departure^2}$$

Substituting:

$$distance = \sqrt{0.040^2 + 0.030^2} = 0.050'$$

Ratio of the Error of Closure

The error of closure is usually expressed as a ratio of the error to the total distance traversed. The ratio is always expressed with the numerator as 1 so the ratio will be 1 foot in X feet.

$$Ratio = \frac{error\ of\ closure}{total\ distance\ traversed} \qquad \text{Equation 14}$$

In our example traverse, the error is 0.05 feet and the total length of the traverse is 606.50 feet. The total length of the traverse lines is not shown in the table in Figure 39 so it is necessary to add all of the distances together. The ratio is:

$$Ratio\ \frac{0.050}{606.50} = 0.000082$$

The error of closure is found by taking the inverse of the ratio (dividing the ratio into 1).

$$\frac{1\ foot}{12,130\ feet}$$

This error closure would be stated as "1 in 12,130". Notice that it is a ratio so it is independent of units. In many jurisdictions the acceptable error of closure is 1 in 10,000 or 1 in 12,000 so the precision of our sample traverse would be acceptable in those jurisdictions. As an aside, an error of 1 in 10,000 equates to 0.01 feet in 100'. This is the smallest graduation on a surveyor's steel measuring tape.

Adjusting the Traverse

The next step is to adjust the traverse. As noted earlier we will use the compass rule for our adjustment. The compass rule distributes the error to latitudes and departures in proportion to the length of the line. Equation 15 shows how this is done. The abbreviation LC is the Latitude Correction. Notice that the total error is shown in the latitude column of our table (0.03'). It is not the error of closure that we calculated from Point 7 to Point 1 (0.05').

$$LC\ Line = \frac{Latitude\ Error * Distance}{Total\ Traverse\ Length} \quad \text{Equation 15}$$

Using the line from Point 1 to Point 2 as an example:

$$LC\ Line\ 1 - 2 = \frac{0.03 * 164.509}{606.502} = 0.008$$

The same procedure is used for departures (DC = Departure Correction).

$$DC\ Line\ 1 - 2 = \frac{0.04 * 164.509}{606.502} = 0.011$$

The adjusted traverse for our example is shown in Figure 40. In order to save space, the latitudes and departures have been reduced to a single column each. North and east values are positive and south and west values are negative.

Traverse Adjusted Table

Point	Bearing	Distance	Latitude	Departure	Lat. Correction	Dep. Correction	Balanced Latitude	Balanced Departure	Corrected Coordinates N	E
1									5000.000	5000.000
	S88° 00' 17"E	164.509	-5.720	164.420	0.008	0.011	-5.728	164.409		
2									4994.272	5164.409
	N00° 22' 20"E	147.746	147.750	0.970	0.007	0.010	147.743	0.960		
3									5142.015	5165.369
	N84° 47' 36"W	77.093	7.000	-76.770	0.004	0.005	6.996	-76.775		
4									5149.011	5088.594
	S83° 10' 02"W	74.504	-8.860	-73.970	0.004	0.005	-8.864	-73.975		
5									5140.147	5014.619
	S03° 34' 20"E	67.024	-66.890	4.180	0.003	0.004	-66.893	4.176		
6									5073.254	5018.795
	S14 23' 25"W	75.626	-73.250	-18.790	0.004	0.005	-73.254	-18.795		
7									5000.000	5000.000
	Sums	606.502	0.030	0.040						

Figure 40 – Adjusted Traverse.

When performing this adjustment, the sign of the error must be taken into consideration. In our example, the ending coordinates are greater than the beginning coordinates in the case of both the north and east coordinates, so the adjustments must be subtracted from the latitudes and departures.

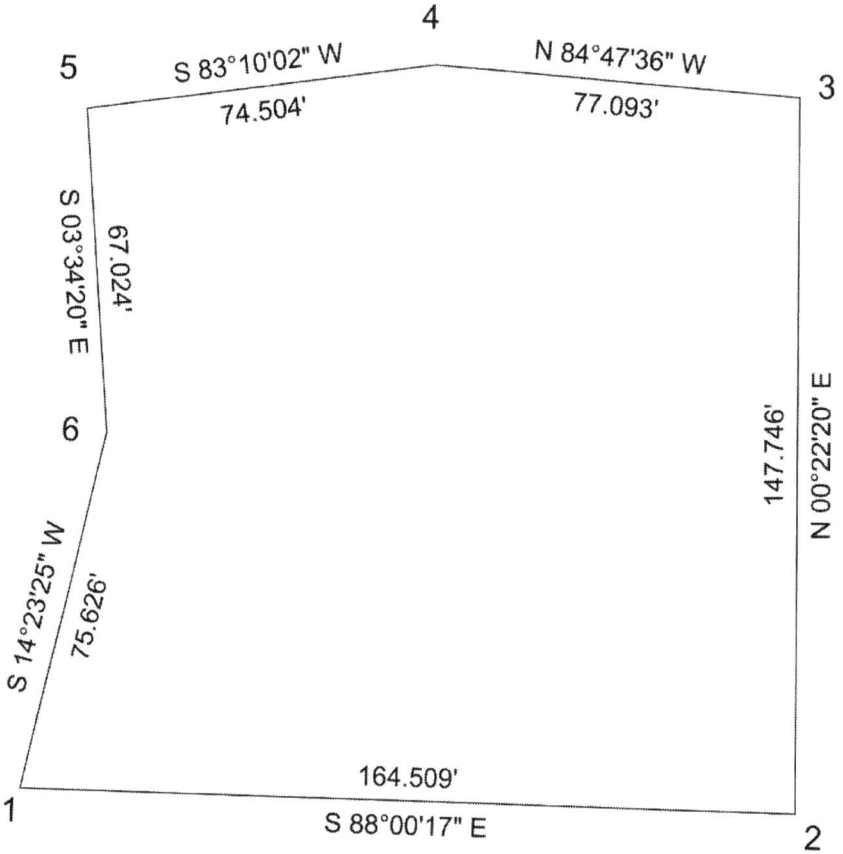

Figure 41 – Adjusted traverse bearings and distances.

Once we have calculated the new latitudes and departures we must calculate new coordinates. We begin with the coordinate of Point 1 in our table in Figure 40 which is N5,000.000, E5,000.000 and by applying the new latitudes and departures we calculate new coordinates for each point. If we have performed our adjustment calculations correctly our last coordinate pair for point 7 will equal our starting coordinate.

Because we have adjusted the coordinates of each point we must calculate a new bearing and distance for each line. This done by inversing between the adjusted coordinates. Our adjusted traverse with the adjusted bearings and distances is shown in Figure 41

Road Geometry

Land surveyors commonly work with road layouts, sometimes called alignments. In this chapter we will consider the mathematics associated with the geometry of roads. We will introduce a number of words that are used by surveyors and civil engineers to describe the features of the geometry typical to all roads. The nomenclature of some of these features can be seen in Figure 42. Most of the nomenclature relates to curves.

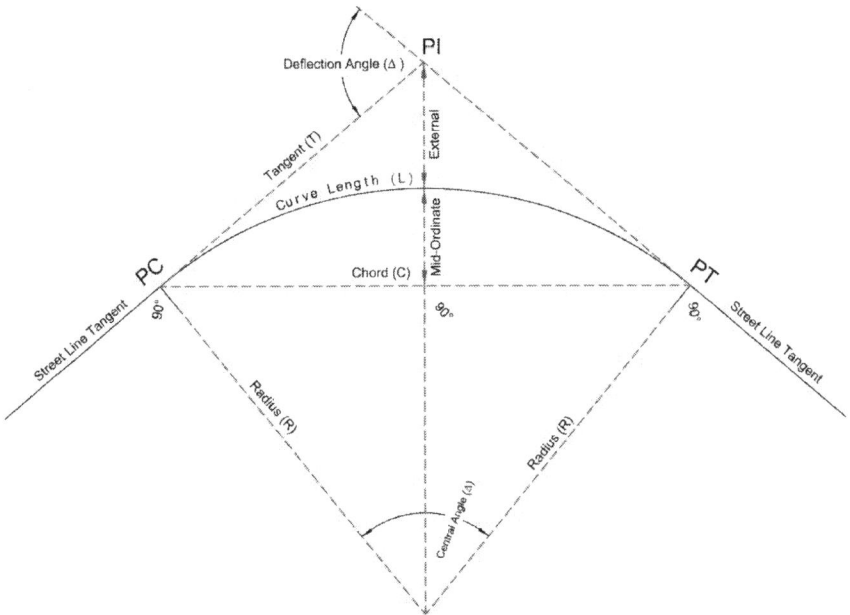

Figure 42 – Road & Curve Nomenclature

Highway and Road Curves

Anyone who has driven down a road knows that all roads have straight sections and curves. Surveyors refer to straight sections as "tangents" and, not surprisingly, curves are called "curves". In most cases when the straight section of a road begins to curve the curve will be "tangent" to the straight section. This is why the straight section is referred to as a tangent. A line is tangent to a curve when it touches the curve at one point. Notice from Figure 43 that, at the point where the straight line touches the curve, it is exactly 90° to a straight line running to the center of the curve – the dashed line in the image. The point where the tangent line touches the curve is called the "Point of Tangency".

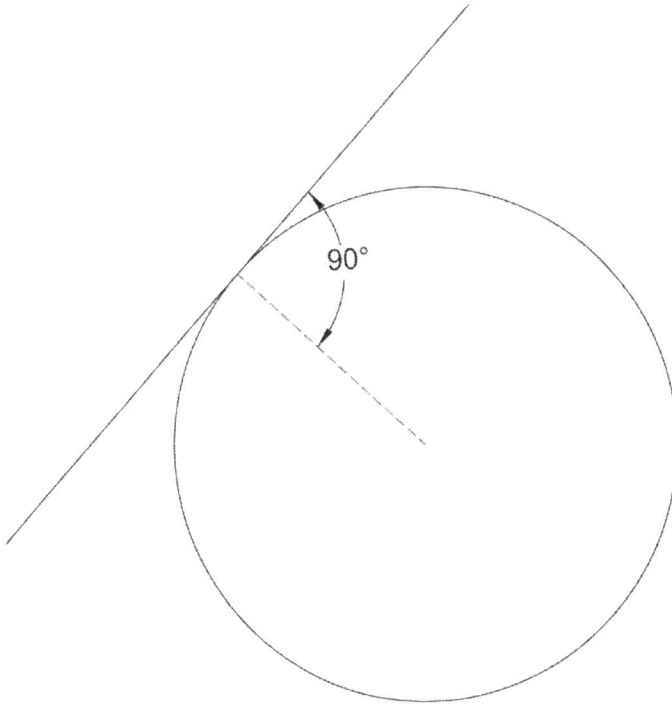

Figure 43 - Line Tangent to Curve.

Let's take a look at a section of a typical road layout as shown in Figure 44.

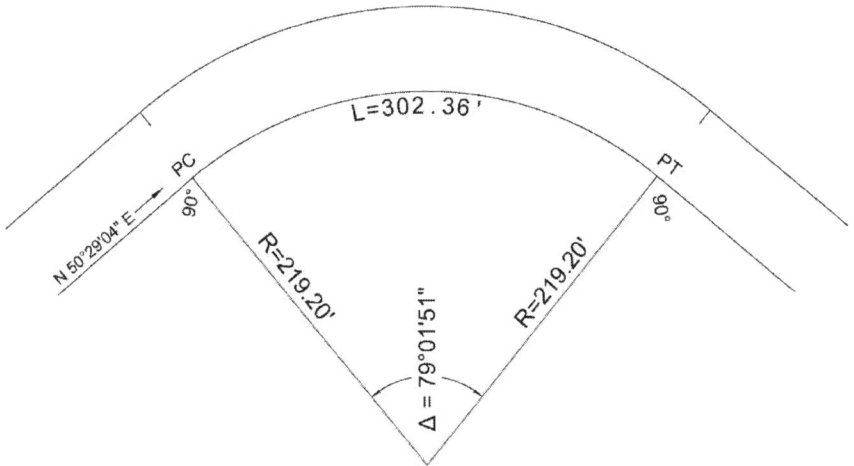

Figure 44 - Radius, Length, Delta

The point at which the straight section of the road begins to curve is labeled "PC". The term PC is an abbreviation for Point of Curvature. The end of the curve, the point where the curve becomes a straight line, is labeled "PT". This stands for Point of Tangency. If you were travelling on the road in the opposite direction, the labels would be reversed. In other words, the PC always comes before PT. In reality, the labels are arbitrary and have nothing to do with the direction that people are driving on the road. The labels actually relate to the direction of the bearings which are used to define the direction of the straight road sections. Notice the arrow on the Bearing N50°29'04"E shows the direction of the straight section leading up to the curve. In this sense, road layouts are like deed descriptions where the description consists of a series of sequential bearings and distances.

Notice in Figure 44 the lines running from the PC and PT intersect at the vertex of the curve. Both of these lines are called the "Radius" of the curve and both are exactly the same length. In the image the length of the radius is 219.20'.

Also, notice in our figure that there is an angle between the two radius lines. This angle is called the "Delta" usually noted by the delta symbol: Δ. In our example the delta is 79°01'51".

The last important item of information in the image is the length of the curve. Curve lengths are usually prefixed using the symbol "L".

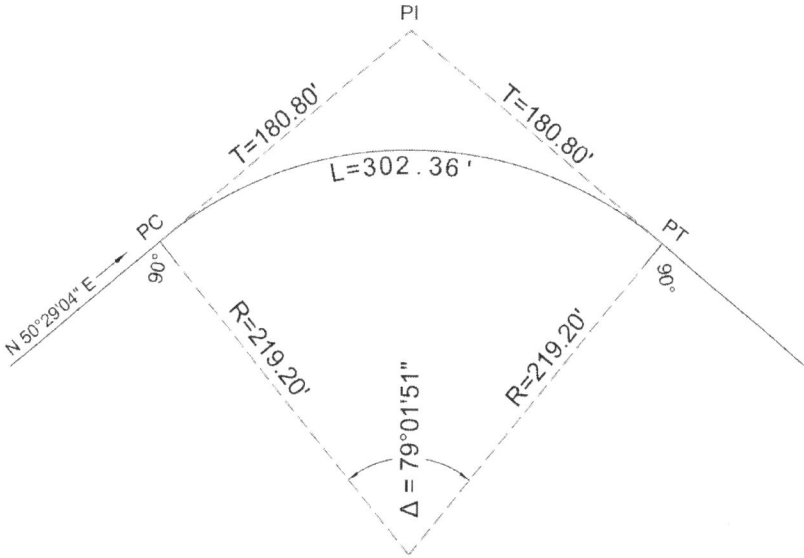

Figure 45 - Tangent

Using the same information as our previous example, Figure 45 shows the addition of the curve Tangents. The curve tangents are simply extensions of the straight road sections to a point where they intersect. This point is labeled the Point of Intersection or "PI".

85

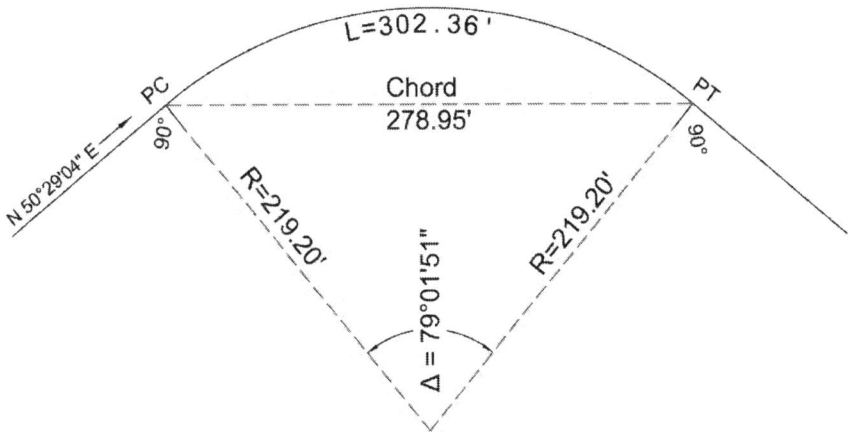

Figure 46 - Chord

The next curve element is shown in Figure 46. This is the Chord. The chord is usually abbreviated "C". The chord is a straight line between the PC and PT. In the old days before computerized data collectors made it easy to calculate points in the field, knowing the chord made it relatively easy to set a PT by setting the instrument on the PC and turning the angle from the road sideline to the PT. It would have also been possible to set up on the PI but this would often place the instrument in the traveled way – which is usually a dangerous place to be.

All of the elements just discussed are shown in Figure 47. Notice that the dimensions of the elements are shown in tabular format at the bottom of the image. On many road layout plans or subdivision plans there is not room to dimension these elements on the lines themselves, so the dimensions are often shown in tabular format.

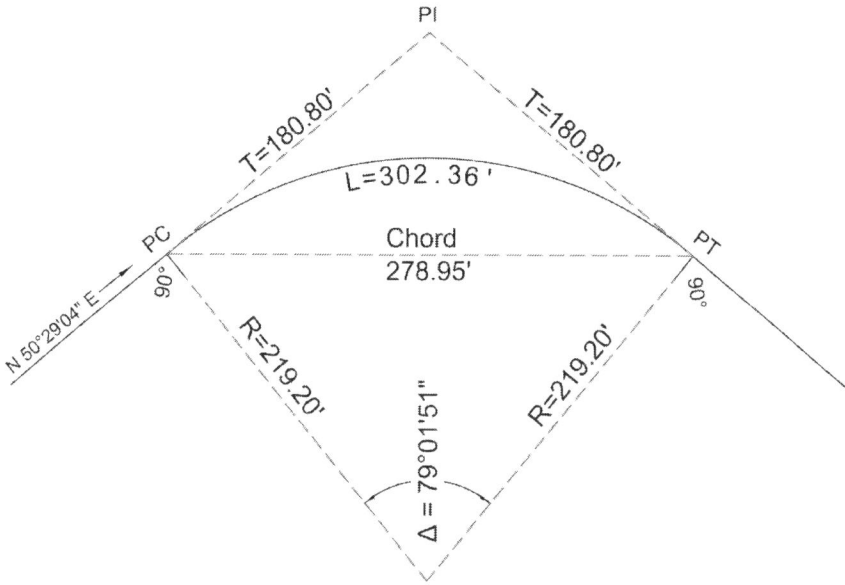

L=302.36'
R=219.20'
△=79°01'51"
C=278.95'
T=180.80'

Figure 47 - Curve Data

Tangent Calculation.

We will now take a look at how the curve elements are calculated. The first element we will calculate will be the tangent. In doing so it is convenient to again resort to our familiar right triangle shown in Figure 48.

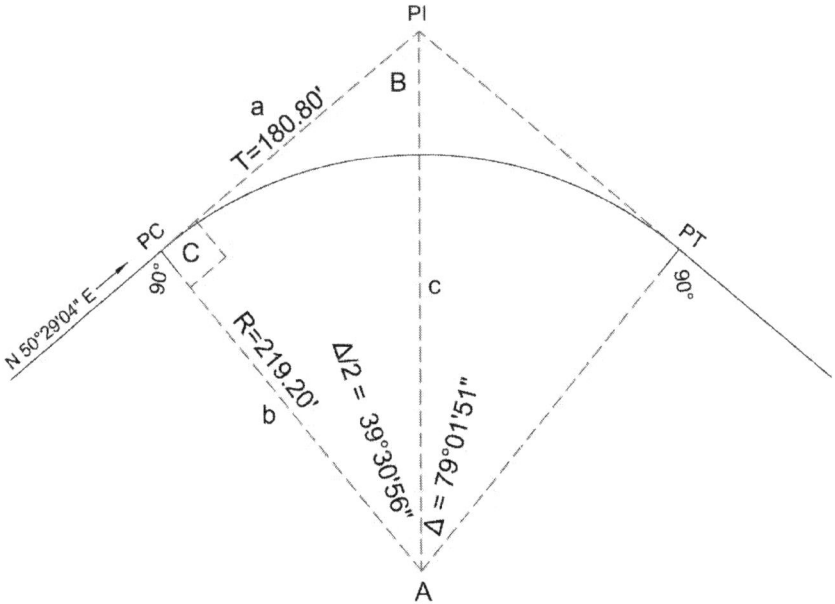

Figure 48 - Tangent Calculation

Notice that the three sides are labeled a, b c and the angles A, B C. Assume that we only know the delta angle, Δ=79°01'51" and the radius, R=219.20'. Because our geometry is symmetrical, we know that angle A must be equal to the delta angle divided by 2. Angle A is therefore equal to 39°30'56". Side b of our triangle is the radius. So, our three variables are A, a, b. We can use Equation 3 to solve the right triangle.

$$tan\,A = \frac{a}{b}\ so\ a = b * \tan A$$

$$a = 219.20' * 0.82479 = 180.80'$$

For a more general solution we know that angle A is $\frac{\Delta}{2}$ so we can write an equation for calculating the tangent as follows:

$$T = R * \tan\frac{\Delta}{2}$$ Equation 16

Calculate the Length of a Curve.

We have seen that the curve length is the length of the arc between the two tangent street side lines. A common way to calculate the arc length is to calculate the circumference of a circle having the same radius as the radius of our curve then simply take the fraction of the circle that the delta represents. We know that a circle contains 360°. If the delta were 90°, for example, the arc length would be ¼ of a circle (90/360 = 0.25). The general equation is:

$$Fraction = \frac{\Delta}{360°}$$ Equation 17

Where the delta (Δ) is in decimal degrees.

The circumference of a circle is given by:

$$C = \pi * 2 * r$$ Equation 18

Where C = the circumference of the circle and r = the radius of the circle. These two equations will allow us to calculate the arc length of any curve as long as we know the delta and radius.

Consider our example in Figure 47. We have a delta angle of 79°01'51" and a radius of 219.20. First, convert our angle to decimal degrees: 79°01'51" = 79.0308°. Next, substitute the values in Equation 17:

$$Fraction = \frac{79.0308°}{360°} = 0.219530$$

Next, we need to know that the circumference of a circle is calculated as:

$$C = \pi * 2 * 219.20' = 1{,}377.274$$

The arc length "L" is calculated by multiplying the Fraction by the circumference:

$$L = 0.219530 * 1{,}377.274 = 302.36$$

We can generalize our equation to calculate the curve length:

$$L = 2 * \pi * R * \frac{\Delta}{360°} \qquad \text{Equation 19}$$

Using our equation to check the above result:

$$L = 2 * \pi * 219.20' * \frac{79.0308°}{360°} = 302.36'$$

By rearranging Equation 19, we can solve for the delta if we know the radius and arc length or we can solve for the radius if we know the delta and arc length.

For example, the equation to calculate the delta would be:

$$\Delta = \frac{L * 360°}{2 * \pi * R} \qquad \textbf{Equation 20}$$

We leave it to the reader to rearrange the equation to solve for the radius.

Deflection Angles

A deflection angle is shown in Figure 42. Deflection angles are sometimes used in road layouts and railroad layouts. It is apparent from the geometry that the deflection angle is equal to the delta angle.

Curves at Intersections.

A very common curve solution can be found at any of the millions of street intersections found in urban and suburban areas throughout the country.

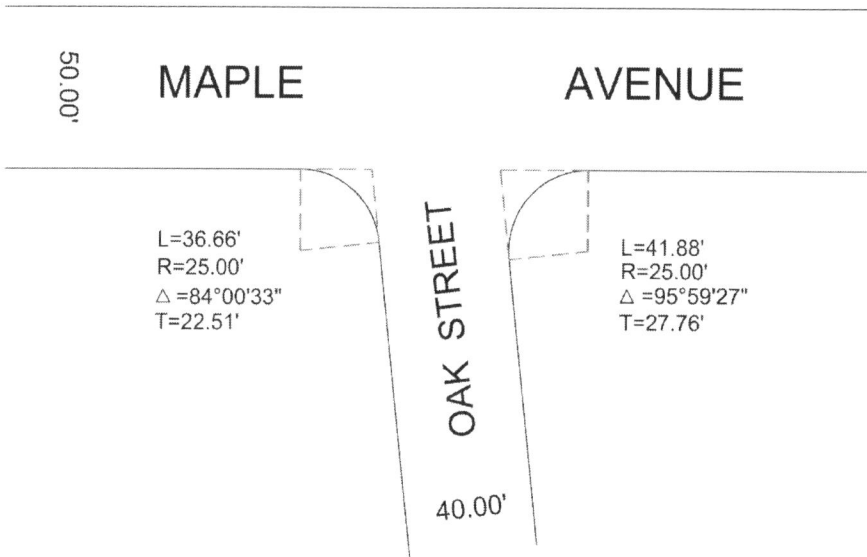

MAPLE   AVENUE

50.00'

L=36.66'
R=25.00'
△ =84°00'33"
T=22.51'

L=41.88'
R=25.00'
△ =95°59'27"
T=27.76'

OAK STREET

40.00'

Figure 49 - Curves at Street Intersection.

Whenever two streets intersect it will usually be necessary to have curves in order to enable vehicles to make the turn from one street onto another. A typical street intersection is shown in Figure 49.

91

Notice that the radius of both curves is the same: 25.00'. However, it is apparent from the delta angle that the streets do not intersect at 90°. This means that the curve radius and tangent will be different for each side of the intersection. The data for each intersection is shown in tabular format for each curve in our image. Notice that the chord is not shown. In order to save space on plans, curve data for street intersections will sometimes only list the radius and length of curve. We have seen that all we need are two quantities in order to be able to calculate the remaining values. So, having the radius and curve length we can calculate the delta and tangent using Equation 16 and Equation 20. On most street layout plans and subdivision plans the street sidelines will have bearings making it easy to calculate delta angles.

Compound Curves.

A compound curve exists when there are two adjacent curves, each having a different radius. Figure 50 shows a compound curve.

In our figure, the curve on the left has a radius of 300.00' and the curve on the right has a radius of 200.00'. The critical thing to understand when working with compound curves is that both curves share a common radius line. The common radius line makes it relatively easy to perform the calculations needed for these curves. Having a common radius line means that the curves are tangent to each other. In our example, the common radius line runs due north (N0°E). The point where the common radius line intersects with the curve (where the radius changes) is called the Point of Compound Curve (PCC).

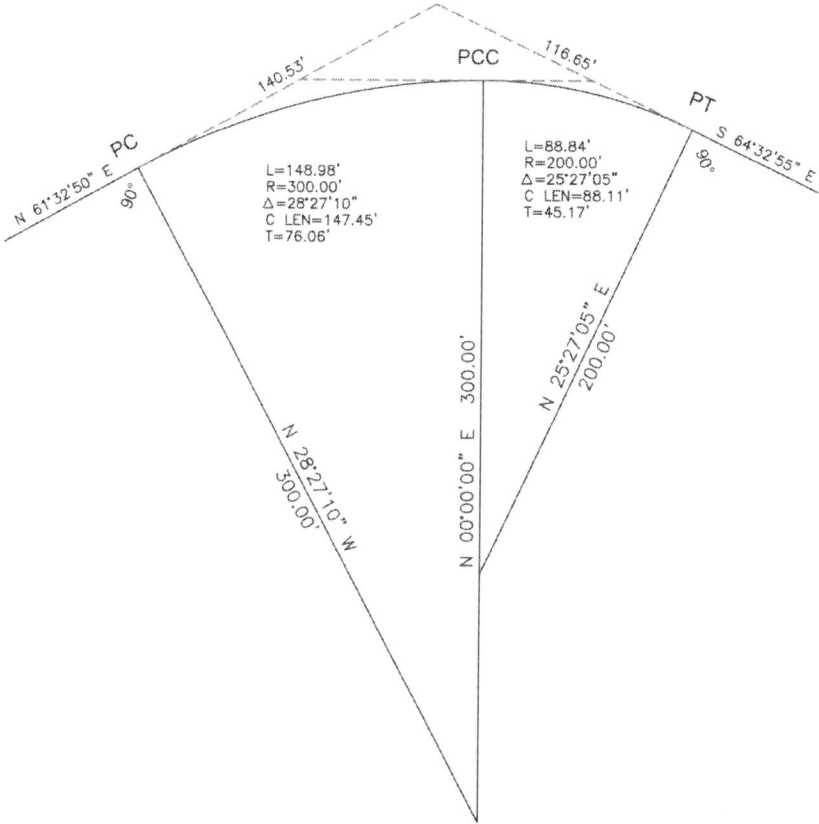

Figure 50 Compound Curve.

Notice in Figure 50 that the intersection of the tangents of the compound curve does not occur opposite the central radius line as it would with a curve having a single radius. This means that the lengths of the tangents are different. One way to calculate the tangent lengths is to use an oblique triangle solution. We have enlarged the relevant area in Figure 51 to make the visualization a bit easier.

93

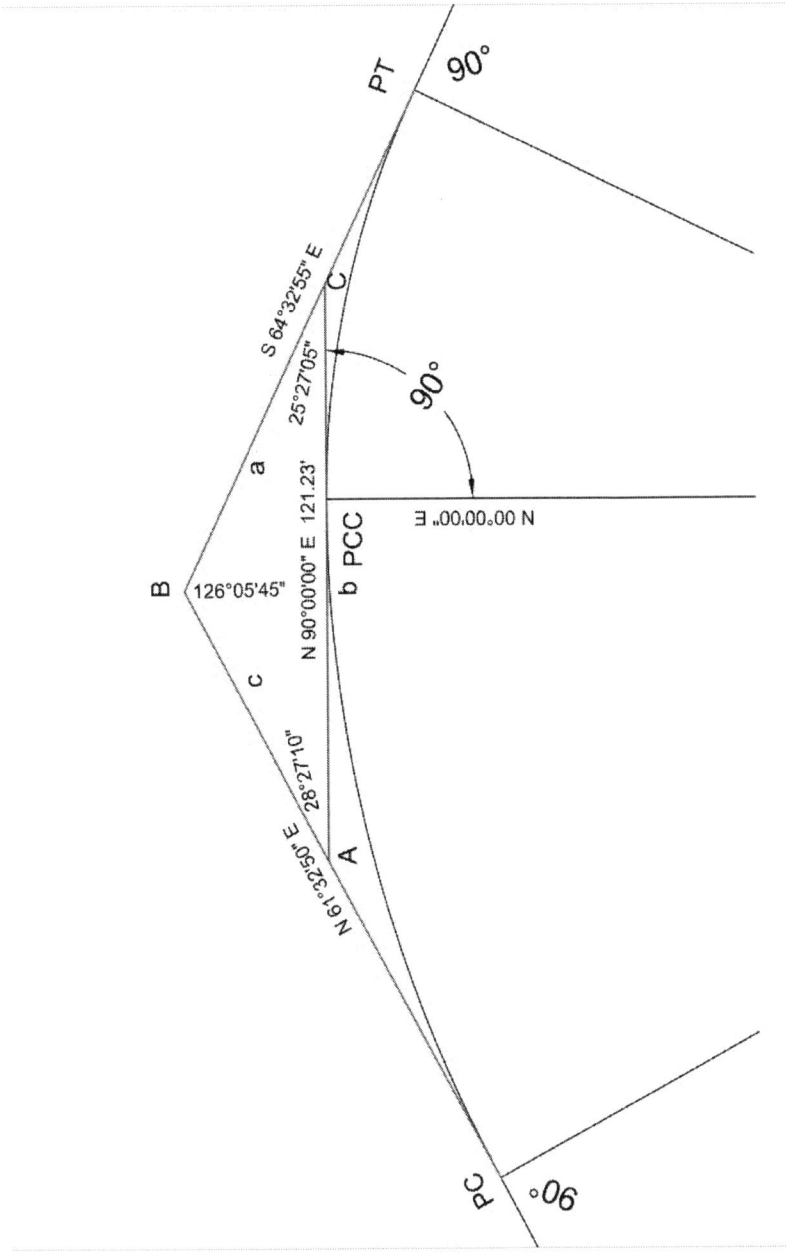

Figure 51 Compound Curve Detail.

Notice in our example, shown in Figure 50 and Figure 51, that we know the bearings of the tangents. These bearings are simply the bearings of the street lines leading into the curve. We also know the bearing of the common radius line (N0°E) and we know that the tangent at PCC (Point of Compound Curve) is 90° to the radius line bearing so we can calculate the bearing of the b side of the triangle (N90°E). Because we have three bearings we can calculate all three interior angles of the oblique triangle. The three angles are shown in Figure 51 and Figure 52.

Another important thing to understand from our curve geometry is that we know the tangent lengths of each of our curves and these tangent lengths can be used to determine the base of the oblique triangle. For example, we can see from Figure 50 that the tangent of the left curve is 76.06' and the tangent of the right curve is 45.17'. If we didn't know these lengths we could easily calculate them. The sum of these tangent lengths is the length of side b of the triangle.

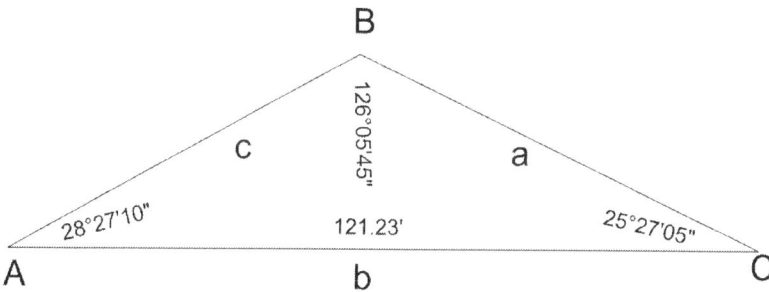

Figure 52 - Compound Curve Triangle

The known quantities are shown in Figure 52. With this information we can use the Equation 6 to calculate the unknown sides of our triangle.

95

$$\frac{a}{\sin A} = \frac{b}{\sin B}$$

Rearrange the equation in order to solve for side a:

$$a = \frac{b}{\sin B} * \sin A$$

Next, we substitute the values and calculate the length of side a.

$$a = \frac{121.23'}{\sin 126°05'45"} * \sin 28°27'10 = 71.48'$$

The total length of the tangent is the sum of the value we just calculated and the value of the tangent of the curve:

$$T = 71.48' + 45.17' = 116.65'$$

We would calculate the remaining tangent using the same method. The values are shown in Figure 50.

Reverse Curves.

In many cases when a road curves to the left then to the right, or vice versa, there is a tangent between the curves, i.e., a straight section between the curves. This is not always the case however. Sometimes a curve leads directly into another curve. This is called a reverse curve. The point where the curve changes direction, is called the PRC (Point of Reverse Curve). A reverse curve is shown in Figure 53. Notice that the bearing of the common radius line is the same. In other words, for a reverse curve to be tangent at the PRC the radius lines of both curves must form a straight line. This can be seen in the image where the bearing of each line, N39°30'56"E, is identical.

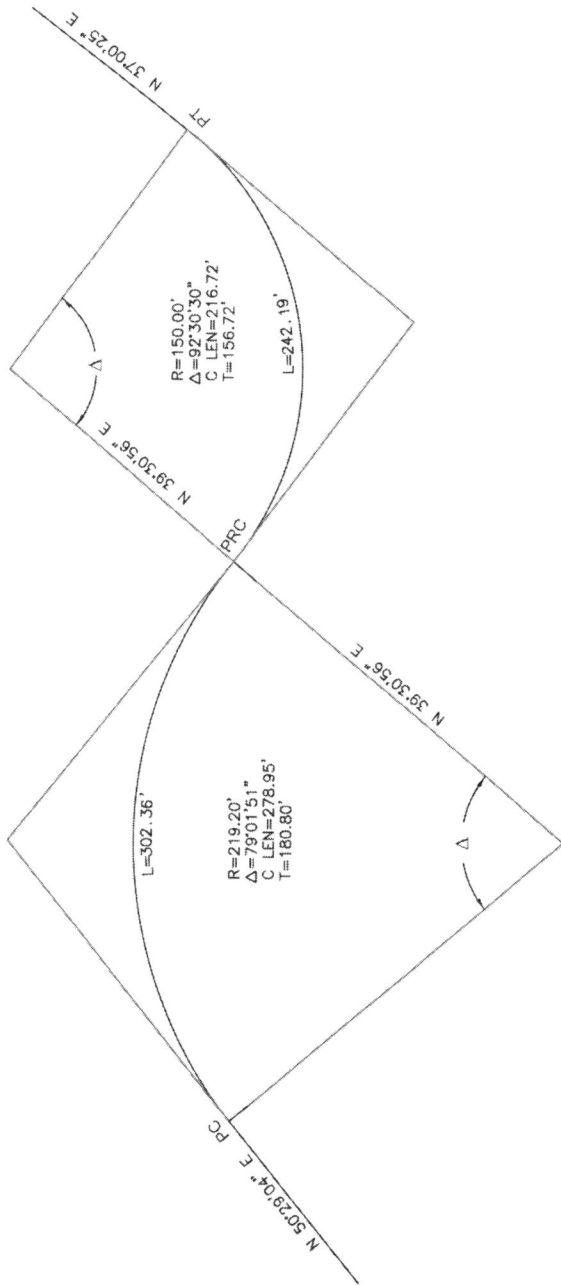

Figure 53 Reverse Curve

97

The calculations for a reverse curve are no different than those we have already discussed.

Non-Tangent Curves

So far in this book we have assumed that whenever a straight section of road enters a curve, the curve is tangent to the straight section. Although this is true the vast majority of the time, this fact should not be taken for granted and one occasionally finds a non-tangent curve.

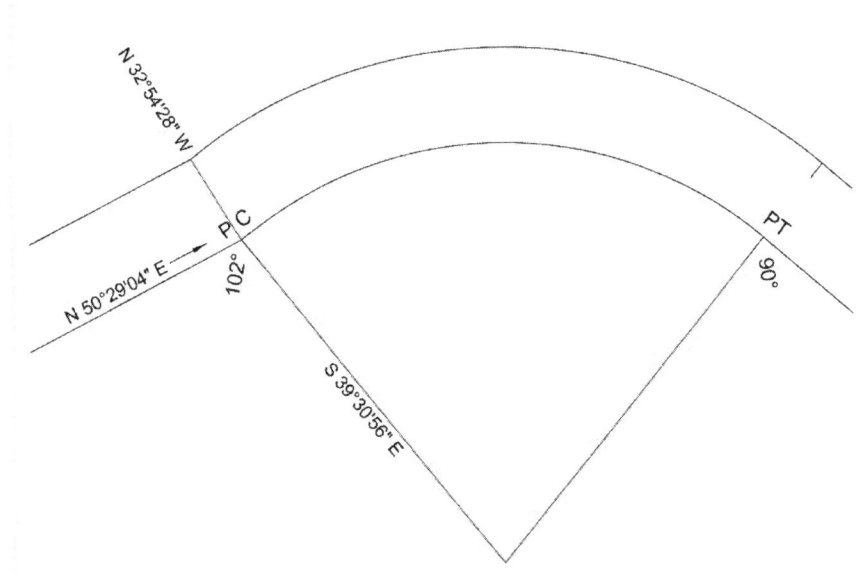

Figure 54 - Non-Tangent Curve

A somewhat extreme example of a curve that is not tangent is shown in Figure 54. It is apparent from the image that the straight section of the road layout is not 90° to the radius line. In fact, it makes an angle of 102° to the radius line. In most cases, when a curve is not tangent to the straight section there will be a notation on the plan to that effect. If the road layout is a constant width, as many, if not most, road layouts are, then the straight section of the road and the curve will not intersect at a projection of the radius line. As can be seen from Figure 54, the bearing of the radius line is different from the bearing of a line connecting the two transition points. In the case of a constant width layout the point opposite the PC (as shown in the image) would be the intersection of the curve with the straight segment.

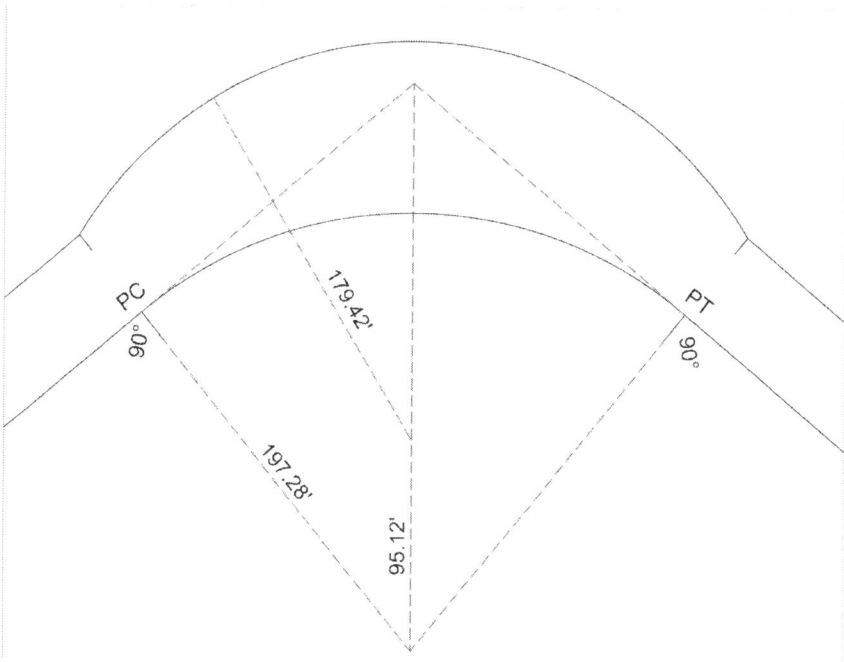

Figure 55 - Non-Tangent Curve

Another example of a non-tangent curve is shown in Figure 55. The lower curve is tangent, but the upper curve is not. In this case the center point of the upper curve is offset 95.12' from the lower curve. Both curves are symmetrical about a line extending from the curve center to the PI.

Road Stations on Centerlines and Baselines.

Road layouts are plans showing surveys of streets, roads and highways that define the extent of the ownership or easements for traveled ways. The ways can be public or private, but most layouts are made by towns, cities counties and states laying out ways for public use. Many such layouts which define the boundaries of the ways include a centerline or baseline.

Layout centerlines and baselines are usually numbered every one-hundred feet with a single integer. These numbers are called Stations. For example, the starting station would be numbered "0". The first station would be numbered "1" which would be 100' from station 0. Station 10 would be 1,000' feet from station 0. So, to get the distance to a particular station in feet, it is necessary to multiply the station number by 100. Intermediate stations are noted with a "+" sign. Figure 56 shows a road layout which has a constant width of 60.00 feet. The +34.45' in this figure at the PC tells us that this point is 2,434.45' from station 0 as measured along the centerline. Notice that the number preceding +34.35 is station 24 which is equal to 2,400 feet.

One must keep in mind that, when a layout contains curves, the centerline distance will not equal the sideline distance. This is apparent in our figure because the curve length of the upper sideline is 246.31' and the lower sideline is 215.53'. The centerline curve length would obviously be midway between these two curve dimensions.

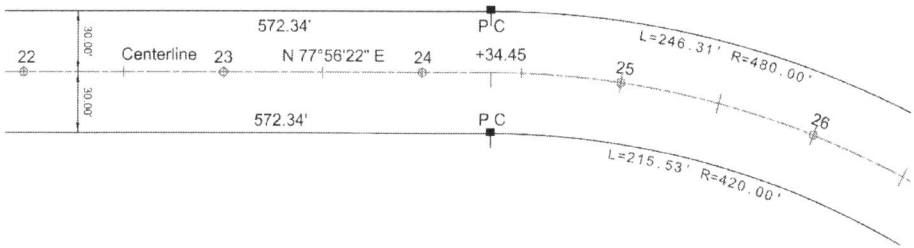

Figure 56 - Road Centerline

Looking again at Figure 56, note that the way is 60.00' in width and the centerline is located in the center of the way, 30.00 feet equidistant from the sidelines. In most road layouts, permanent monuments are set at the PCs and PTs on the sidelines of the layout and at intermediate points when the straight tangents are very long, for example at least every 500 feet. Notice the small black squares at the PC of the layout in the image representing monuments. Road layout monuments set by government authorities are usually stone or concrete.

For the layout shown in Figure 56, the relationship between the centerline and the monuments at the PC is apparent. The monuments are located 30.00' either side of station 24+34.45 at 90° to the centerline bearing N77°56'22"E. So, using a centerline is a very straightforward and convenient way to dimension a road layout.

Some road layouts do not use a centerline. Some use a baseline as in Figure 57. Notice that the baseline is not centered between the road sidelines. Baselines are often derived from the actual traverse that the surveyor ran when the road layout was made. For example, at the early stages of a street design the surveyor may not know where the street will be located exactly so a randomly located traverse line is used in the field which later becomes a baseline. In other cases, narrow roads with heavy traffic would make it very dangerous for a surveyor to occupy and make

101

measurements on a centerline so it may be necessary to run lines outside of the traveled way.

When a baseline is used, it is necessary to mathematically tie the baseline to the layout in some way. This is usually done using "tie lines". Two tie lines are shown in Figure 57 to tie station 24 to the two monuments located at the PC. The tie lines have bearings and distances from station 24 to the PCs. Because the baseline is randomly located, each tie line will be different. In some cases, it is possible for the tie lines to be 90° to the baseline but in that case a separate station would be required for each PC. (In our example a tie line at 90° could be used for the lower PC but not for the upper PC unless one of the baselines were extended.) In Figure 58 there are 90° offsets to monuments from the baseline running along Sycamore St.

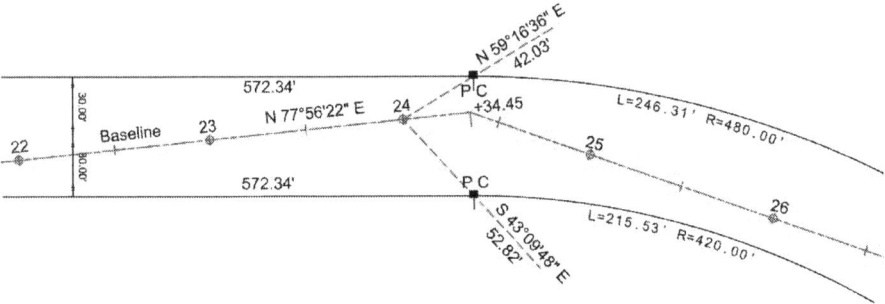

Figure 57 - Road Baseline

As we have discussed, calculating the distances between stations is straightforward. For example, consider Figure 58. calculate the distance between station 29+11.03 and station 28+8.26.

$$Station\ 29 + 11.03 = (29 * 100') + 11.03' = 2,911.03'$$

$$Station\ 28 + 8.26 = (28 * 100') + 8.26 = 2,808.26'$$

$$Distance = 2{,}911.03' - 2{,}808.26' = 102.77'$$

Intersecting Centerlines and Baselines

A street intersection is shown in Figure 58. June Street contains a centerline and Sycamore Street contains a baseline. The intersection of the centerline and baseline occurs at Sycamore St. station 28+43.70 and at June Street station 79+29.22. Also notice the angle point in the Sycamore Street baseline at 29+11.03.

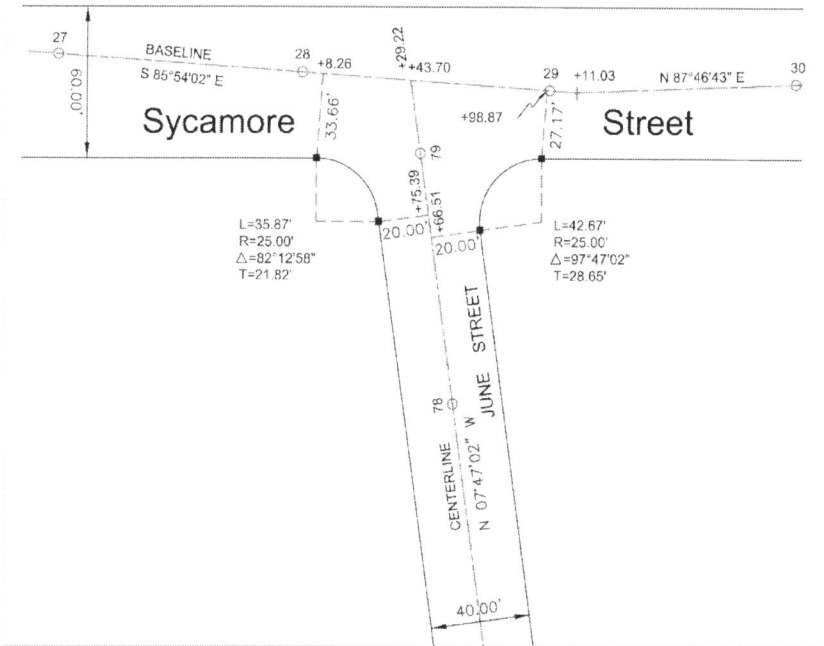

Figure 58 Baselines and Centerlines

Notice that all of the ties to the PCs and PTs are at 90° to the baseline and centerline and each tie has its own station.

103

Calculating Area

Calculating area is a very common and fundamental procedure that surveyors have been engaged in for centuries. Land is often valued, at least in part, by its area. Other things being equal, the larger the area of a parcel the greater its economic value. Area is measure of the "quantity" of land contained within its boundaries. Area is always measured in the horizontal plane. The area of a 100 foot square parcel of land on a steep Rocky Mountain slope is identical to the 100 foot square measured on a flat Iowa cornfield.

In the United States, area is commonly measured in square feet or acres. One acre = 43,560 square feet. This is one surveying measurement that every aspiring surveyor should memorize.

In the days before computers and calculators were widely available, measuring area of an irregular parcel was often a laborious and complex process such as using double meridian distances (DMD) (beyond our scope here). Sometimes parcels of land would be broken down into a series of triangles. The area of each triangle would then be calculated, and all of the triangle areas could then be added together. The area of a triangle is simply the length of its base times it height divided by 2, although it is often more convenient to use another method to calculate area, which we will discuss.

It was common to resort to graphic methods for calculating areas such as using a planimeter. A planimeter is a small mechanical device that has an arm which can be dragged around the perimeter of a parcel of land drawn on paper. This resulted in number and a scale factor from which a rough area could be determined. Because some of these methods were crude, at least by modern standards, the areas found in old deed descriptions and plans is sometimes rather inaccurate.

Calculating the area of rectangular parcels.

Calculating the area of a rectangular parcel of land is simple and precise. Recall that a rectangle, by definition, has 90° interior angles. The area of a rectangle is simply its width times it height.

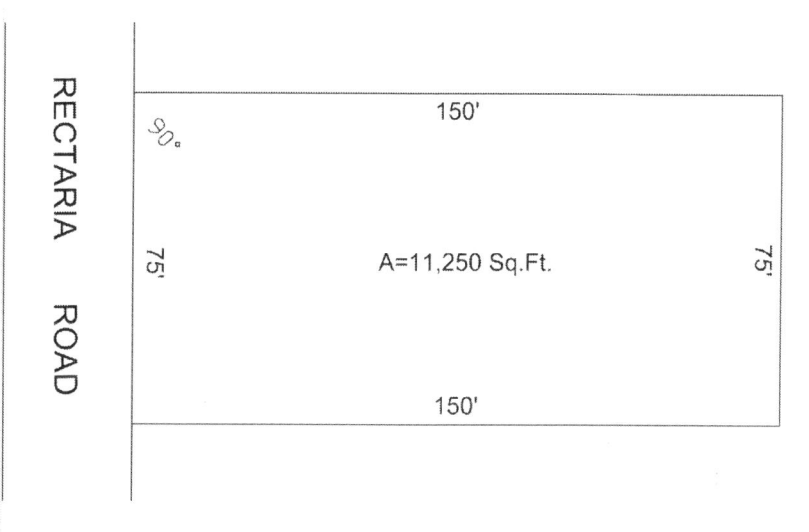

Figure 59 Area of a Rectangular Parcel.

To calculate the area of the parcel in Figure 59, we first notice that the all of the interior angles are 90°. The area of a rectangle is given by:

$$A = L * W = 75 * 100 = 11,250 \, Sq. Ft.$$

If we want to express the area in acres we would divide the area in square feet by 43,560.

$$A = \frac{11,250 \ sq.ft.}{43,560} = 0.258 \ Acres$$

A parcel of land lying at the intersection of two streets will usually have a radiused corner. A typical lot like this is shown in Figure 60. The lot is rectangular, but the area will be less than a rectangle because the curve removes some of the area from one corner. There are a number of ways to calculate the reduced area. One way is illustrated in our figure.

We will proceed to calculate the area by dividing the parcel into rectangles. The area inside the circular sector will be calculated as a portion of a circle.

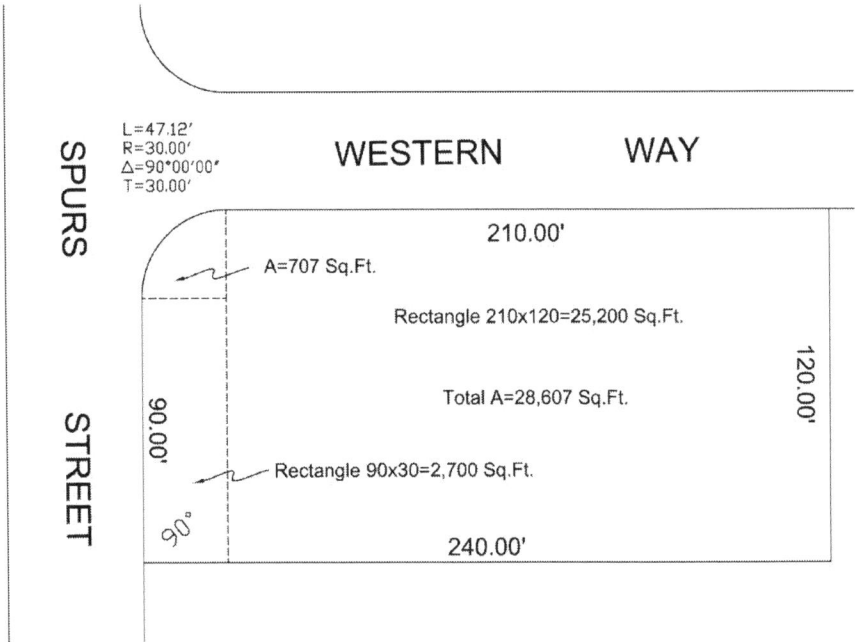

Figure 60 Area of Corner Lot

The radius of the curve in the figure is 30.00 feet. With this in mind, we can simply remove a 30' wide strip of land from the left edge of the parcel thereby creating a rectangle 210' x 120'. The area of this rectangular section is simply:

$$A = 210 * 120 = 25{,}200 \; sq.ft.$$

This leaves us with a rectangle running along Spur St. The area of this rectangle is:

$$A = 90 * 30 = 2{,}700 \; sq.ft.$$

The last area to calculate is the area of the circular sector. This is calculated as a fraction of a circle. In this case the Delta is 90° so we have ¼ of a circle. The area of a circle is given by:

$$A = \pi r^2$$

So, for a full circle with a radius of 30.00' we calculate the area as:

$$A = \pi * 25.00^2 = 2{,}827.43 \; sq.ft.$$

The area of ¼ of the circle is:

$$A = 2.827.43 * 0.25 = 706.86 \; sq.ft.$$

If the Delta were not 90° we would simply use a ratio of the Delta to 360° as we did when we calculated the curve length in the section on road geometry.

Summing the areas, we have:

$$Total \; A = 25{,}200 + 2{,}700 + 707 = 28{,}607 \; sq.ft.$$

Calculating Area using Triangles

Many parcels of land have more complex shapes than simple rectangles, so we need to find another method for calculating the

107

area of these parcels. If the parcel under consideration does not contain curves we can use triangles to calculate the area.

In order to use triangles, we must know the lengths of each side of every triangle. We can use the following equations (Heron's formula) to calculate the area of a triangle.

$$Area = \sqrt{s(s-a)(s-b)(s-c)}$$ **Equation 21**

$$s = \frac{1}{2}(a+b+c)$$ **Equation 22**

Where a, b and c are the lengths of the sides of the triangle.

An example of a four-sided parcel of land is shown in **Figure 61.** We have drawn a line from corner to corner in order to create two triangles. So, how do we calculate the distance of this line?

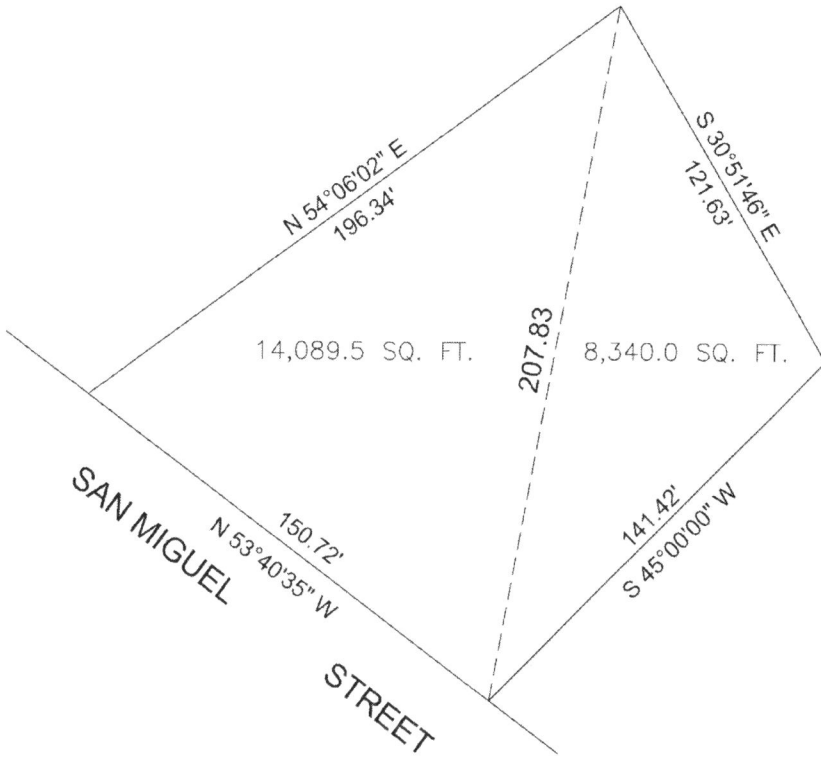

Figure 61 Using Triangles to Calculate Area.

One way would be to use coordinate geometry to obtain coordinates for each corner of the parcel then inverse to obtain the distance of the dashed line shown in the image. Assume that we did this and determined that the length of the line is 207.83'. We now have all of the information that we need to calculate the areas of the triangles.

We can use Equation 21 and Equation 22 to calculate the areas.

Triangle 1:

$$s = \frac{1}{2}(a + b + c) = \frac{1}{2}(150.72 + 196.34 + 207.83) = 277.445$$

$$A = \sqrt{277.445(277.445 - 150.72)(277.445 - 196.34)(277.445 - 207.83)}$$
$$= 14{,}089.5 \; sq.\,ft.$$

The same procedure would be used to calculate the area of the second triangle, and the two would be added together to arrive at our answer.

For larger parcels, there can be many triangles to solve. One such parcel is shown in Figure 62. As you can see, it would be rather time-consuming calculating all of these triangles. Fortunately, most surveying coordinate geometry software will do the hard work for you. However, learning the process that we have discussed would allow you to calculate almost any area with a simple hand calculator. You could also quite easily set up a spreadsheet to do the calculations and just input the three distances for each triangle into three columns designed for that purpose.

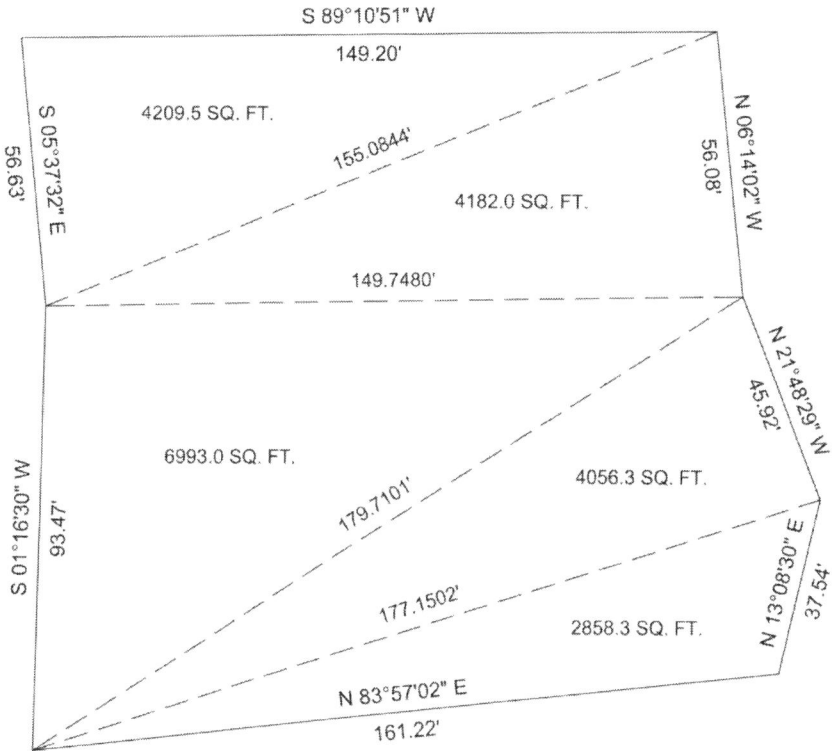

Figure 62 Area of Irregular Parcel using Triangles.

111

Printed in Great Britain
by Amazon